W9-ADU-239

Lecture Notes in Mathematics

A collection of informal reports and seminars
Edited by A. Dold, Heidelberg and B. Eckmann, Zürich

297

John Garnett
University of California, Los Angeles, CA/USA

Analytic Capacity and Measure

Springer-Verlag
Berlin · Heidelberg · New York 1972

AMS Subject Classifications (1970): 30 A 86, 28 A 10, 30 A 82

ISBN 3-540-06073-1 Springer-Verlag Berlin · Heidelberg · New York
ISBN 0-387-06073-1 Springer-Verlag New York · Heidelberg · Berlin

This work is subject to copyright. All rights are reserved, whether the whole or part of the material is concerned, specifically those of translation, reprinting, re-use of illustrations, broadcasting, reproduction by photocopying machine or similar means, and storage in data banks.

Under § 54 of the German Copyright Law where copies are made for other than private use, a fee is payable to the publisher, the amount of the fee to be determined by agreement with the publisher.

© by Springer-Verlag Berlin · Heidelberg 1972. Library of Congress Catalog Card Number 72-93416. Printed in Germany.

Offsetdruck: Julius Beltz, Hemsbach/Bergstr.

MATH.-SCI.
QA
331
.G3
1340343
Math Sci
Sep.

CONTENTS

INTRODUCTION

Let Ω be an open subset of the complex plane. We are interested in two classes of analytic functions on Ω. The first, denoted by $H^\infty(\Omega)$, is the set of bounded analytic functions on Ω; and the second, which we call $A(\Omega)$, is the set of functions in $H^\infty(\Omega)$ possessing continuous extensions to the Riemann sphere S^2. Because of the maximum principle, it is often more convenient to discuss $H^\infty(\Omega)$ and $A(\Omega)$ in terms of the complementary set $E = S^2 \setminus \Omega$ instead of Ω. Rotating S^2 so that $\infty \in \Omega$, we can assume that E is a compact plane set.

Perhaps the best way to describe the problems considered below is to prove two elementary theorems. Let E be a compact plane set and let $\Omega = S^2 \setminus E$ be its complement.

<u>Painlevé's Theorem</u>: Assume that for every $\varepsilon > 0$, the set E can be covered by discs the sum of whose radii does not exceed ε. Then $H^\infty(\Omega)$ consists only of constants.

<u>Proof</u>: For each $\varepsilon > 0$ we cover E by a collection of discs, the sum of whose radii does not exceed ε, and we let Γ_ε be the boundary of the union D_ε of these discs. If $f \in H^\infty(\Omega)$ and $f(\infty) = 0$, then by Cauchy's theorem

$$f(z) = \frac{1}{2\pi i} \int_{\Gamma_\varepsilon} \frac{f(\zeta)d\zeta}{z - \zeta}, \quad z \notin \bar{D}_\varepsilon .$$

Thus $|f(z)| \le \dfrac{\varepsilon \cdot \|f\|}{2\pi \operatorname{dist}(z,\Gamma_\varepsilon)}$. Sending ε to 0 we get $f(z) = 0$ for all $z \in \Omega$.

Theorem: If E has positive area, then $A(\Omega)$ contains non-constant functions.

Proof: Write

$$F(z) = \iint_E \frac{d\xi d\eta}{\zeta - z} \qquad \zeta = \xi + i\eta .$$

Clearly $F(z)$ is analytic on the complement Ω of E, and, being the convolution of the locally integrable function $-1/\zeta$ and the characteristic function of a bounded set, $F(z)$ is continuous on the complex plane. Since $\lim_{z\to\infty} F(z) = 0$, F is in $A(\Omega)$, and F is not constant since $\lim_{z\to\infty} zF(z) = -\text{Area } E$.

That proves the theorem, but let us examine the function F more closely. At infinity $F(z)$ has the expansion

$$F(z) = -\text{Area}(E)/z + a_2/z^2 + \cdots .$$

Define R by $\pi R^2 = \text{Area}(E)$ and let Δ be the disc $\{\zeta : |\zeta - z| \leq R\}$. Then

$$|F(z)| \leq \iint_E \frac{d\xi d\eta}{|\zeta - z|} = \iint_{E\cap\Delta} \frac{d\xi d\eta}{|\zeta - z|} + \iint_{E\setminus\Delta} \frac{d\xi d\eta}{|\zeta - z|} .$$

Since E and Δ have the same area, so do $E\setminus\Delta$ and $\Delta\setminus E$, while the integrand is larger on $\Delta\setminus E$ than it is on $E\setminus\Delta$. Thus

$$\int_{E\setminus\Delta} \frac{d\xi d\eta}{|\zeta - z|} \leq \int_{\Delta\setminus E} \frac{d\xi d\eta}{|\zeta - z|}$$

and so

$$|F(z)| \leq \iint_{\Delta} \frac{d\xi d\eta}{|\zeta - z|} = 2\pi R .$$

This gives us a function $g(z) = F/2\pi R$ in $H^{\infty}(\Omega)$ such that $\|g\| \leq 1$ and

$$g(z) = b_1/z + b_2/z^2 + \cdots$$

where $|b_1| \geq \frac{R}{2} = \frac{1}{2}\sqrt{\frac{\text{Area}(E)}{\pi}}$. In other words, we have estimated the analytic capacity of E (defined in Chapter I below) in terms of the area of E.

The hypothesis of each sample theorem is measure theoretic (in Painlevé's theorem the measure is one dimensional Hausdorff measure), and each is proved by representing a function as the Cauchy integral of a Borel measure. Our purpose is to survey what can be said concerning two problems:

1°. Representing functions in $H^{\infty}(\Omega)$ as Cauchy-Stieltjes integrals

$$f(z) = \int_E \frac{d\mu(\zeta)}{\zeta - z} .$$

2°. Estimating or describing analytic capacity in terms of measures, and applying such estimates to approximation problems.

These notes contain much that is old and a little that is new. Hopefully, they are intelligible to the graduate student who knows elementary real and complex analysis and a little functional analysis, and who is interested in analytic capacity and related fields. As three fine expositions of rational approximation theory [28], [81], [88] are already in print, it seems unnecessary to discuss that theory in any depth. Thus the Melnikov-Vitushkin estimates on line integrals have been ignored, and when we use Vitushkin's approximation techniques in Chapter V we give suitable references but no details.

In some instances a theorem has a person's name attached to it, often simply because that is what the theorem is called. But no doubt some important results have not been ascribed to their originals, and unattributed theorems should not be assumed the author's discovery.

Throughout each chapter there are exercises and problems. Some exercises are very routine, and some problems are old and famous, but the only real distinction is that I think I can do the exercises.

Chapter I is an exposition of the theory of analytic capacity. It begins at the beginning, and thus has some overlap with other sources. Chapter II concerns the Cauchy integral representation. It contains a simple characterization of Cauchy transforms. The relation between bounded analytic functions, Hausdorff measure, and Newtonian potential theory is taken up in Chapter III. In Chapter IV we discuss three examples, and in Chapter V applications are made to approximation theory.

I wish to thank H. Alexander, A. Davie, L. Hedberg, P. Koosis, K. Pietz, H. Royden, J. Wermer and L. Zalcman for valuable suggestions and conversations. I am especially grateful to T. Gamelin for advice at every stage of the preparation of this paper. I also thank Laurie Beerman for typing the manuscript.

Certain notation should be mentioned. $\Delta(z,\delta)$ stands for the open disc $\{\zeta : |\zeta - z| < \delta\}$ and $S(z,\delta)$ is the closed square of side

δ and center z. Its sides are parallel to the axes. The symbol μ denotes a finite complex Borel measure; S_μ is its support and $|\mu|$ is its variation measure. Unless otherwise indicated, a.e. refers to area, and $\|f\|$ is the supremum of $|f|$ over its domain. C_0^∞ are the compactly supported infinitely differentiable functions, and $C_0^\infty(D)$ are those with support inside D. A function g is in L^p_{loc} if $|g|^p$ is integrable over every compact set. Finally

$$\frac{\partial}{\partial \bar{z}} = \frac{1}{2}\left(\frac{\partial}{\partial x} + i\,\frac{\partial}{\partial y}\right) .$$

CHAPTER I. ANALYTIC CAPACITY

§1. Basic Properties

Let E be a compact plane set and let $\Omega = S^2 \setminus E$. When f is analytic on Ω its derivative at ∞ is computed using the local coordinate $1/z$, so that

$$f'(\infty) = \lim_{z \to \infty} z(f(z) - f(\infty)) .$$

Expanding f in a Taylor series about ∞,

$$f(z) = a_0 + a_1/z + a_2/z^2 + \cdots$$

we have $f'(\infty) = a_1$. In other words,

$$f'(\infty) = \frac{1}{2\pi i} \int_\Gamma f(\zeta) d\zeta$$

whenever the curve Γ separates E from ∞. Define the <u>analytic capacity</u> and <u>continuous analytic capacity</u> respectively as follows

$$\gamma(E) = \sup\{|f'(\infty)| : f \in H^\infty(\Omega),\ \|f\| \leq 1\}$$

$$\alpha(E) = \sup\{|f'(\infty)| : f \in A(\Omega),\ \|f\| \leq 1\} .$$

<u>Theorem 1.1</u>: $H^\infty(\Omega)$ consists only of the constants if and only if $\gamma(E) = 0$; $A(\Omega)$ consists only of the constants if and only if $\alpha(E) = 0$.

<u>Proof</u>: Clearly, if $\gamma(E) > 0$ then $H^\infty(\Omega)$ contains non-constant functions.

On the other hand, if $H^\infty(\Omega)$ is not trivial there is $f \in H^\infty(\Omega)$ with $f(\infty) = 0$ and $f(z_0) \neq 0$ for some $z_0 \neq \infty$. Then the function $(f(z) - f(z_0))/(z_0 - z)$ is in $H^\infty(\Omega)$ and has derivative $f(z_0)$ at ∞, so that $\gamma(E) > 0$. The same argument shows $A(\Omega)$ is nontrivial if and only if $\alpha(E) > 0$.

If $f \in H^\infty(\Omega)$ and $\|f\| \leq 1$, then

$$g(z) = \frac{f(z) - f(\infty)}{1 - \overline{f(\infty)}f(z)}$$

is in $H^\infty(\Omega)$, $\|g\| \leq 1$, $g(\infty) = 0$ and

$$g'(\infty) = \frac{f'(\infty)}{1 - |f(\infty)|^2} .$$

Thus when computing the extremum $\gamma(E)$, we can restrict our attention to functions vanishing at ∞. Using the standard notation

$$A(E,M) = \{f \in H^\infty(\Omega) : \|f\| \leq M,\ f(\infty) = 0\}$$

we have

$$\gamma(E) = \sup\{|f'(\infty)| : f \in A(E,1)\} .$$

Similarly, letting

$$C(E,M) = A(E,M) \cap A(\Omega) ,$$

we have

$$\alpha(E) = \sup\{|f'(\infty)| : f \in C(E,1)\} .$$

If $f(\infty) = 0$, then $\lim_{z\to\infty} zf(az + b) = af'(\infty)$, and we have the invariance properties

$$\gamma(aE + b) = |a|\gamma(E)$$

$$\alpha(aE + b) = |a|\alpha(E) .$$

It is clear from the definitions that γ and α are monotone: $\gamma(E) \leq \gamma(F)$, $\alpha(E) \leq \alpha(F)$ if $E \subset F$. It is also clear that $\alpha(E) \leq \gamma(E)$. However these two quantities are not commensurate. For, $\gamma(E)$ depends only on the unbounded component of Ω: $\gamma(E) = \gamma(\hat{E})$ where $\hat{E}$ is the union of E and the bounded components of Ω. So if E is the circle $\{z : |z - a| = r\}$, then $\gamma(E) > 0$, but $\alpha(E) = 0$ by Morera's theorem. Another example is obtained by taking E to be an interval on the real axis. Then $\alpha(E) = 0$, again by Morera's theorem, while $\gamma(E) > 0$ because Ω can be mapped conformally onto the unit disc.

When E is connected, so that Ω is simply connected, the class $A(E,1)$ arises in a well known proof of the Riemann mapping theorem [2, p. 222]. Indeed, we have

<u>Theorem 1.2</u>: Assume that E is connected but not a point. Let g be the conformal map of Ω onto the unit disc satisfying $g(\infty) = 0$, $g'(\infty) > 0$. Then $\gamma(E) = g'(\infty)$.

<u>Proof</u>: Since $g \in A(E,1)$, we have $g'(\infty) \leq \gamma(E)$. Let $f \in A(E,1)$. Applying Schwarz's lemma to $F = f \circ g^{-1}$, we have $|F'(0)| \leq 1$. But $F'(0) = f'(\infty)/g'(\infty)$ so that $|f'(\infty)| \leq g'(\infty)$. Therefore $\gamma(E) \leq g'(\infty)$.

Consequently, we see that if E is the disc $\{|z - a| \leq \delta\}$, then $\gamma(E) = \delta$; and the extremal function is $\delta/(z-a)$. And if E is a line segment of length ℓ, then $\gamma(E) = \ell/4$. For this it is enough to take $E = [-2,2]$ and observe that the conformal map $g : S^2 \setminus E \to \Delta(0,1)$ satisfies

$$g^{-1}(w) = w + 1/w .$$

We can now estimate analytic capacity in terms of diameters as follows

Corollary 1.3: For any set E

$$\alpha(E) \leq \gamma(E) \leq \mathrm{diam}(E) .$$

If E is connected, then

$$\gamma(E) \geq \frac{\mathrm{diam}(E)}{4} .$$

Proof: The first assertion follows by monotonicity, because E lies in a disc of radius $\mathrm{diam}(E)$.

To prove the second assertion we can assume that E is not a point.

Let $g(z) = \gamma(E)/z + a_2/z^2 + \cdots$ be the Riemann map of the unbounded component of Ω onto the unit disc. Fix $z_0 \in E$ and write

$$f(w) = \frac{\gamma(E)}{g^{-1}(w) - z_0} .$$

Then f is univalent on $|w| < 1$, $f(0) = 0$, and $f'(0) = 1$. By the Koebe-Bieberbach theorem [23, p. 279], the range of f contains $|z| < 1/4$, so that if $z_1 \in E$, we have

$$\frac{\gamma(E)}{|z_1 - z_0|} \geq 1/4 .$$

Corollary 1.3 implies that E is totally disconnected when $\gamma(E) = 0$.

The estimate on analytic capacity given in Corollary 1.3 is sharp in the case of a line segment. In the introduction we gave another estimate:

$$\gamma(E) \geq \alpha(E) \geq \frac{1}{2} \sqrt{\frac{\text{Area}(E)}{\pi}} .$$

In Chapter III this will be improved by a factor of 2, so that it is sharp in the case of a disc.

When E is not compact define

$$\gamma(E) = \sup\{\gamma(K) : K \text{ compact}, K \subset E\}$$

$$\alpha(E) = \sup\{\alpha(K) : K \text{ compact}, K \subset E\} .$$

It is then clear that $\gamma(U) = \alpha(U)$ for all open sets U. A normal family argument shows

$$\gamma(E) = \inf\{\alpha(U) : U \text{ open}, U \supset E\}$$

when E is compact.

The condition $\gamma(E) = 0$ is necessary and sufficient for the set E to be removable for bounded analytic functions.

Theorem 1.4: Let E be a relatively closed subset of an open set U and assume $\gamma(E) = 0$. If $f \in H^\infty(U \setminus E)$, then f has a unique extension in $H^\infty(U)$.

Proof: The uniqueness is trivial because E is nowhere dense. Let $z_0 \in E$. Since by 1.3 E is totally disconnected, there is an analytic simple closed curve Γ in $U\setminus E$ which encloses z_0. Let D be the domain bounded by Γ. Using the Cauchy integral we can write $f = f_1 + f_2$ in a neighborhood of Γ, where $f_1 \in H^\infty(D)$, and $f_2 \in H^\infty(S^2\setminus(E \cap D))$. Since $\gamma(E \cap D) = 0$, f_2 is constant, and f extends analtyically to D.

The same result is true if $f \in A(U\setminus E)$ and γ is replaced by α. However the above argument only works if E is a compact subset of U. The simplest proof of the full result uses Vitushkin's localization operator ([28] II, 1.7) and would be a digression at this point.

Theorem 1.5: If E is a relatively closed subset of an open set U and $\alpha(E) = 0$, then every $f \in A(U\setminus E)$ has a unique extension in $A(U)$.

The Semi-additivity Problem: Show there exists a constant C such that

$$\gamma(E_1 \cup E_2) \leq C(\gamma(E_1) + \gamma(E_2))$$

for some reasonable class of sets (like the Borel sets). Equivalent formulations of this quite important conjecture are given in [18] and [81]. It is not known whether there is a constant C such that

$$\gamma(E_1 \cup E_2) \leq C\gamma(E_1)$$

for all compact E_2 such that $\gamma(E_2) = 0$. The same problems for the continuous analytic capacity α are also open and from the point of view of rational approximation theory the important question is whether or not

$$\alpha(E_1 \cup E_2) \leq C\alpha(E_1)$$

when E_2 is compact and $\alpha(E_2) = 0$.

Exercise 1.6: Prove that if $\{E_n\}$ is a decreasing sequence of compact sets and $E = \bigcap E_n$, then $\gamma(E) = \lim \gamma(E_n)$. Prove that if $\{F_n\}$ is a sequence of compact sets such that $\gamma(F_n) = 0$, then $\gamma(\bigcup F_n) = 0$. Determine if the above two assertions hold with γ replaced by α.

Exercise 1.7: In §6 it is proved that any subset of $[0,1]$ with positive inner Lebesgue measure has positive analytic capacity. Use this fact and the usual construction of a non-measurable set to exhibit subsets $\{E_n\}$ of $[0,1]$ such that $\gamma(E_n) = 0$ for all n but $\bigcup E_n = [0,1]$. Use the same ideas to construct two sets E_1 and E_2 such that $E_1 \cup E_2 = [0,1]$, but $\gamma(E_1) = \gamma(E_2) = 0$.

§2. Schwarz's Lemma

As is suggested by the proof of Theorem 1.2, there is a close connection between analytic capacity and the Schwarz lemma. Indeed we have the following inequalities which, though elementary, are the reasons that the extremal quantity $\alpha(E)$ is so important in approximation theory.

Theorem 2.1: Let $f \in A(E,1)$. Then for $z_0 \in \Omega$ we have

$$|f(z_0)| \leq \frac{\gamma(E)}{\operatorname{dist}(z_0,E)} .$$

If f has a zero of multiplicity n at ∞, and E has diameter d, then

$$|f(z_0)| \leq \frac{d^{n-1}\gamma(E)}{(\operatorname{dist}(z_0,E))^n} .$$

The same inequalities hold if $f \in C(E,1)$ and $\gamma(E)$ is replaced by $\alpha(E)$.

Proof: Consider

$$g(z) = \frac{1}{z - z_0} \cdot \frac{f(z) - f(z_0)}{1 - \overline{f(z_0)}f(z)} .$$

Then $g \in H^\infty(\Omega)$, $g(\infty) = 0$, and $|g(z)| \leq \frac{1}{\operatorname{dist}(z_0,E)}$. Thus

$$|f(z_0)| = |g'(\infty)| \leq \frac{\gamma(E)}{\operatorname{dist}(z_0,E)} .$$

If f has order n at ∞ and $z_1 \in E$ with $|z_1 - z_0| = \operatorname{dist}(z_0,E)$, then

$$h(z) = \left(\frac{z - z_1}{d}\right)^{n-1} f(z)$$

is in $A(E,1)$, so that

$$|f(z_0)| \leq \left(\frac{d}{|z - z_0|}\right)^{n-1} |h(z_0)| \leq \frac{d^{n-1}\gamma(E)}{(\operatorname{dist}(z_0,E))^n} .$$

§3. Two Classical Theorems

For any domain Ω the Hardy space $H^p(\Omega)$, $0 < p < \infty$, is the class of analytic functions h on Ω such that the subharmonic function

$|h(z)|^p$ has a harmonic majorant. This definition is conformally invariant and coincides with the usual definition [50] when Ω is the unit disc.

Assume that $\infty \in \Omega$ and that $\partial\Omega = \Gamma$ consists of finitely many pairwise disjoint analytic Jordan curves $\Gamma_1, \ldots, \Gamma_n$. We need two classical theorems, Fatou's theorem and the F. and M. Riesz theorem. Rather than give complete proofs of these quite well known results we merely derive them from their counterparts for the unit disc proved for instance in [25] or [50]. We will use the fact that the conformal map of a simply connected domain D to the unit disc extends analytically across any analytic arc on ∂D. See [2, pp. 224-227]. The symbols ds and $|d\zeta|$ denote the arc length element on Γ.

Fatou's Theorem: If $h(z) \in H^p(\Omega)$, $1 \leq p < \infty$, then $h(z)$ has a non-tangential boundary values $h(\zeta)$ at almost every ζ on Γ, $h(\zeta)$ is in $L^p(ds)$, and if $h(\infty) = 0$,

$$h(z) = \frac{-1}{2\pi i} \int_\Gamma \frac{h(\zeta) d\zeta}{\zeta - z} .$$

Here "almost every" refers of course to arc length.

Proof: For each $\zeta \in \Gamma$ take a neighborhood V of ζ such that $V \cap \Omega$ is simply connected and $V \cap \partial\Omega$ is an analytic arc. Then the conformal mapping of $V \cap \Omega$ to the unit disc Δ extends to be analytic across $V \cap \partial\Omega$. Since (the restriction of) h is in $H^p(\Omega \cap V)$, and that class is conformally invariant, we can use the Fatou theorem for Δ and the boundary properties of the conformal mapping to obtain a boundary function $h(\zeta)$ in L^p on $V \cap \partial\Omega$. The boundary functions for different choices of V coincide on the intersections because a nonzero function in $H^p(\Delta)$

cannot vanish on a set of positive measure. Thus there is a well defined boundary function $h(\zeta)$ in $L^p(ds)$.

If $\{\Gamma_t\}$ is a system of curves in Ω converging to Γ as t tends to 0, and if W is a disc contained in one of the neighborhoods V chosen above; then the measures $h(z)dz$ on the $\Gamma_t \cap W$ converge weak star to $h(\zeta)d\zeta$ on $\Gamma \cap W$, because the corresponding assertion holds in the unit disc. A partition of unity argument now shows that the measures $h(z)dz$ on Γ_t converge to $h(\zeta)d\zeta$ weak star. Hence if $h(\infty) = 0$ and $z_0 \in \Omega$,

$$h(z_0) = \frac{-1}{2\pi i} \int_\Gamma \frac{h(\zeta)d\zeta}{\zeta - z_0} .$$

The "boundary value" part of the above theorem holds as well if $0 < p < 1$; but the corresponding Cauchy integral may not exist. Naturally we now regard $H^p(\Omega)$ as a space of functions on Γ even though they are defined only almost everywhere.

A measure μ <u>annihilates</u> $A(\Omega)$ if $\int f d\mu = 0$ for all $f \in A(\Omega)$.

<u>F. and M. Riesz Theorem</u>: Every measure μ on Γ which annihilates $A(\Omega)$ is of the form

$$\mu = \frac{h(\zeta)d\zeta}{\zeta^2}$$

where $h(z) \in H^1(\Omega)$.

<u>Proof</u>: First consider the case where Ω is simply connected. The conformal mapping τ from Ω to the disc $\Delta = \{|w| < 1\}$ sending ∞ to 0 is analytic across $\partial\Omega$, and a homeomorphism of $\overline{\Omega}$ onto $\overline{\Delta}$.

So if g is in $A(\Delta)$ then $g \circ \tau$ is in $A(\Omega)$, and the functional on $C(\partial\Delta)$

$$L(g) = \int g \circ \tau \, d\mu$$

annihilates $A(\Delta)$. Hence there is $H(w) \in H^1(\Delta)$ such that $L(g) = \int g(w)H(w)dw$. It follows that for $f \in C(\partial\Omega)$,

$$\int f(z)d\mu(z) = \int f(\tau^{-1}(w))H(w)dw$$

$$= \int f(z)H(\tau(z))\tau'(z)dz \ .$$

Since $\tau'(z)$ vanishes twice at ∞ and is bounded on Ω, $h(z) = z^2 H(\tau(z))$ is in $H^1(\Omega)$ and

$$\mu = \frac{h(z)dz}{z^2} \ .$$

For the general case write $\mu = \Sigma \mu_j$ where μ_j has support the curve Γ_j, and for $z \notin \Gamma_j$ let $g_j(z)$ be the Cauchy transform of μ_j:

$$g_j(z) = \int_{\Gamma_j} \frac{d\mu_j(\zeta)}{\zeta - z} \ .$$

Write $\Omega = \bigcap \Omega_j$, where $\partial\Omega_j = \Gamma_j$. If $f \in A(\Omega_j)$ vanishes at ∞, and $k \neq j$, then

$$\int_{\Gamma_j} f(z)g_k(z)dz = \int_{\Gamma_j} f(z) \int_{\Gamma_k} \frac{d\mu_k(\zeta)}{\zeta - z} dz = 2\pi i \int_{\Gamma_k} f(\zeta)d\mu_k(\zeta) \ .$$

Writing

$$\nu_j = \mu_j + \frac{1}{2\pi i} \sum_{k \neq j} g_k(\zeta)d\zeta$$

on Γ_j, we have $\int f d\nu_j = \int f d\mu = 0$, so that by the simply connected case

$$\nu_j = \frac{G_j(\zeta)d\zeta}{\zeta}$$

with $G_j \in H^1(\Omega_j)$. Now for $z \in \Omega$,

$$g_j(z) = \int_{\Gamma_j} \frac{d\nu_j(\zeta)}{\zeta - z} - \sum_{k \neq j} \frac{1}{2\pi i} \int_{\Gamma_j} \frac{g_k(\zeta)d\zeta}{\zeta - z} .$$

Every integral in the summation vanishes, so that by Fatou's theorem

$$g_j(z) = -2\pi i \frac{G_j(z)}{z} .$$

It now follows from the definition of ν_j that $\mu_j = g(\zeta)d\zeta$ on Γ_j where

$$g(\zeta) = \frac{-1}{2\pi i} \sum_k g_k(\zeta)$$

is in $H^1(\Omega)$. Since $g(\infty) = 0$ and $\lim_{z\to\infty} zg(z) = \int d\mu = 0$, we get the desired representation taking $h(z) = z^2 g(z) \in H^1(\Omega)$.

Exercise 3.1: Prove, directly from the definition, that if $f,g \in H^2(\Omega)$, then $fg \in H^1(\Omega)$.

Exercise 3.2: Let Ω be a domain bounded by finitely many pairwise disjoint analytic Jordan curves. Let μ be a complex measure such that

$$\hat{\mu}(z) = \int_{\partial\Omega} \frac{d\mu(\zeta)}{\zeta - z} \qquad z \in \Omega$$

is in $H^1(\Omega)$. Prove $\mu \ll ds$. Hint: consider first the case $\hat{\mu} = 0$ on Ω.

§4. The Ahlfors Function

Suppose E is a compact plane set whose complement $\Omega = S^2 \setminus E$ is connected. A normal families argument shows there is an extremal function $f \in A(E,1)$ with $f'(\infty) = \gamma(E)$. When E is connected, Theorem 1.2 tells us there is a unique such extremal function; moreover it is a univalent mapping of Ω to the unit disc. It turns out that for any E there is a unique extremal function, called the Ahlfors function, and if E has n components, the Ahlfors function is an n-fold covering map onto the unit disc. We will first prove that an extremal function is a covering map when Ω is finitely connected, and in the next section we use the machinery developed to easily derive the uniqueness for any E. If Ω is finitely connected, then by the Riemann mapping theorem (applied n times) we can assume $\partial\Omega$ consists of analytic Jordan curves. In that case one gets a stronger assertion.

Theorem 4.1: (Ahlfors [1]). Let Ω be a domain such that $\infty \in \Omega$ and $\partial\Omega$ consists of pairwise disjoint analytic Jordan curves $\Gamma_1, \ldots, \Gamma_n$. Let $E = S^2 \setminus \Omega$. Then

(a) $\alpha(E) = \gamma(E)$.

If $f \in A(E,1)$ is an extremal function, then

(b) f is analytic across $\partial\Omega$

(c) $|f| = 1$ on $\partial\Omega$

(d) f has n zeros on Ω.

Of course, the argument principle then easily yields that for $|w| < 1$, $f(z) = w$ has n solutions (counting multiplicities). Proofs of this theorem occur for example in [1], [14], [32] and [60], and the approach taken in [60] is quite elementary. The argument given here is similar to the proof in [14] but we begin with the Hahn-Banach theorem instead of a variational argument.

Proof: Let σ be a measure on $\partial\Omega$ of norm $\|\sigma\| = \alpha(E)$ such that $\int f d\sigma = f'(\infty)$ for all $f \in A(\Omega)$. The measure

$$\mu = \frac{dz}{2\pi i} - \sigma$$

is orthogonal to $A(\Omega)$, so that by the F. and M. Riesz Theorem

$$\sigma = \psi(z)dz$$

where $\psi(z) = 1/2\pi i + b_2/z^2 + \cdots$ is in $H^1(\Omega)$. The function ψ solves the dual extremal problem

$$\int_{\partial\Omega} |\psi(\zeta)|ds = \inf\left\{\int_{\partial\Omega} |h(\zeta)|ds,\ h \in H^1(\Omega),\ h(\infty) = \frac{1}{2\pi i}\right\}$$

and is called the Garabedian function.

If $g \in A(E,1)$, then $g \cdot \psi \in H^1(\Omega)$ and $|g\psi| \le |\psi|$ on $\partial\Omega$. Therefore

$$|g'(\infty)| = \left|\int_{\partial\Omega} g(\zeta)\psi(\zeta)d\zeta\right| \le \int_{\partial\Omega} |\psi| ds = \alpha(E) ,$$

and taking $g'(\infty) = \gamma(E)$ we obtain $\gamma(E) \le \alpha(E)$, so that (a) holds.

Now let $f \in A(E,1)$ satisfying $f'(\infty) = \gamma(E)$. Then

$$\int_{\partial\Omega} f(\zeta)\psi(\zeta)d\zeta = \gamma(E) = \int_{\partial\Omega} |\psi(\zeta)| ds .$$

This implies

(i) $f(\zeta)\psi(\zeta)d\zeta \ge 0$ on $\partial\Omega$,

and because ψ cannot vanish on a set of positive measure,

(ii) $|f(\zeta)| = 1$ a.e. on $\partial\Omega$.

Condition (i) and the analyticity of each Γ_j imply that $f(z)\psi(z)$ is analytic across $\partial\Omega$. For if $\zeta(w)$ is analytic on $r < |w| < 1/r$ with $\zeta(|w| = 1) = \Gamma_j$, then (for r near 1), $g(w) = f(\zeta)\psi(\zeta)$ is in H^1 of the annulus $1 < |w| < 1/r$ and $g(w)\zeta'(w) \ge 0$ on $|w| = 1$. Schwarz reflection and Morera's theorem can be used to extend $g(w)\zeta'(w)$ to the larger annulus $r < |w| < 1/r$, so that $f(\zeta)\psi(\zeta)$ is analytic on $\partial\Omega$. Now the argument principle and condition (i) imply that

(iii) $f(z)\psi(z)$ has n zeros on Ω.

For each $\zeta \in \Gamma$ take a neighborhood V of ζ such that $V \cap \Omega$ is simply connected, $V \cap \Gamma$ is an arc and $f(z)\psi(z)$ has no zeros on $V \cap \Omega$. After a conformal map $z = z(w)$ we have nonvanishing functions $\tilde{f}(w)$ and $\tilde{\psi}(w)$ on the unit disc Δ and an arc J in $\partial\Delta$ (corresponding to $V \cap \Gamma$) on which $\tilde{f}(w)\tilde{\psi}(w)$ is continuous. Write $\tilde{f} = S_1F_1$, $\tilde{\psi} = S_2F_2$ where S_j is a singular function and F_j is outer [50]. Then as $\tilde{f}\tilde{\psi}$ is analytic across J, the same holds for S_1S_2 and thus for S_1. Since

$$F_1(w) = \exp \frac{1}{2\pi} \int_{\partial\Delta} \frac{e^{i\theta} + w}{e^{i\theta} - w} \log|F_1(e^{i\theta})|d\theta ,$$

and $|F_1| = 1$ on J by (ii), we see that F_1 is analytic across J as well. Therefore $\tilde{f}$ is analytic across J and $|\tilde{f}| = 1$ on J. Going back to V, we obtain (b) and (c).

Finally, (b) and (c) tell us that f has at least n zeros on Ω while (iii) ensures that there can be no more than n zeros.

The proof of 4.1 yields some additional information on the extremal function ψ. We know that ψ is analytic on $\partial\Omega$, because f is, and that ψ has no zeros on Ω, by (c) and (iii). Moreover $\arg f$ has variation -2π around each Γ_j, and $f(\zeta)\psi(\zeta)d\zeta \geq 0$, so that $\arg \psi$ has variation zero around each Γ_j. Hence $\log \psi$ is single valued on Ω.

Corollary 4.2: When $\partial\Omega$ consists of finitely many pairwise disjoint analytic Jordan curves the Garabedian function $\psi(z)$ is analytic across $\partial\Omega$ and $\log \psi(z)$ is single valued on Ω.

The proof of 4.1 applies as well to other extremal problems [4], [59], [71]. For example, let $z_0 \neq \infty$ and let $g \in A(E,1)$ maximize $|g(z_0)|$. Then g satisfies (b), (c) and (d) of 4.1. The solutions of the two extremal problems coincide when Ω is simply connected, but not otherwise.

The Hardy space $H^2(\Omega)$, when given the norm $\|g\|^2 = \int |g|^2 ds$, is a subspace of $L^2(ds)$ on which evaluations of points of Ω are continuous. Hence there is a unique function (the Szego kernel function) $K(\zeta,\infty) \in H^2(\Omega)$ such that

$$g(\infty) = \int_{\partial\Omega} g(\zeta)\overline{K(\zeta,\infty)}ds, \qquad g \in H^2(\Omega) .$$

Therefore

$$\sup\{|g(\infty)| : g \in H^2(\Omega),\ \|g\|_2 \leq 1\} = \|K(\zeta,\infty)\|$$

$$= (K(\infty,\infty))^{1/2},$$

and the extremal function for this problem is

$$\frac{K(\zeta,\infty)}{(K(\infty,\infty))^{1/2}} .$$

Theorem 4.3: Let Ω be an unbounded domain such that $E = \partial\Omega$ consists of finitely many analytic Jordan curves $\Gamma_1,\ldots,\Gamma_n$. Let $K(\zeta,\infty)$ be the Szego kernel for ∞ and let $\psi(z)$ be the Garabedian function. Then

$$K(\zeta,\infty) = \frac{1}{2\pi\gamma(E)}\,(2\pi i\psi(\zeta))^{1/2}$$

where the root is chosen so that $K(\infty,\infty) = 1/2\pi\gamma(E)$. Moreover,

$$\sup\{|g(\infty)| : g \in H^2(\Omega),\ \|g\| \leq 1\} = \frac{1}{\sqrt{2\pi\gamma(E)}} .$$

Proof: By an elementary Hilbert space argument $K(\zeta,\infty)/K(\infty,\infty)$ is the unique element of minimal norm in the convex set

$$S = \{g \in H^2(\Omega) : g(\infty) = 1\},$$

and its norm is

$$\left\|\frac{K(\zeta,\infty)}{K(\infty,\infty)}\right\| = \frac{1}{(K(\infty,\infty))^{1/2}} .$$

Let $g \in S$. Since $(\psi(\zeta))^{1/2}$ is in H^2, we have by Fatou's theorem and Exercise 3.1

$$1 = g(\infty) = \frac{1}{2\pi i} \int_{\partial\Omega} g(\zeta) \frac{f(\zeta)}{\gamma(E)} (2\pi i \psi(\zeta))^{1/2} \, d\zeta ,$$

so that

$$1 \leq \|g\|_2 \frac{1}{2\pi\gamma(E)} \left(\int |2\pi\psi(\zeta)| \, ds \right)^{1/2}$$

$$= \frac{\|g\|_2}{\sqrt{2\pi\gamma(E)}} .$$

Hence $\|g\| \geq \sqrt{2\pi\gamma(E)}$, and equality holds for $g(\zeta) = (2\pi i \psi(\zeta))^{1/2}$. This means $K(\infty,\infty) = \dfrac{1}{2\pi\gamma(E)}$ and $K(\zeta,\infty) = \dfrac{(2\pi i \psi(\zeta))^{1/2}}{2\pi\gamma(E)}$.

Theorem 4.3 gives an explicit method for computing $\gamma(E)$. The rational functions with poles off $\overline{\Omega}$ are dense in $H^2(\Omega)$, [72]. Using the Gram-Schmidt process, one can extract an orthonormal basis $\{u_n\}_{n=1}^{\infty}$ for $H^2(\Omega)$. Then since

$$K(\zeta,\infty) = \sum_{n=1}^{\infty} \overline{u_n(\infty)} \cdot u_n(\zeta)$$

$$\gamma(E) = \frac{1}{2\pi} \left(\sum |u_n(\infty)|^2 \right)^{-1} .$$

However in many cases the computations involved are too unweildy to yield substantial information about $\gamma(E)$.

Exercise 4.4: Let E be a compact plane set and let f be an Ahlfors function for E. Show that the finite zeros of f lie in the convex hull of E. See [66].

§5. Uniqueness of the Ahlfors Function and Havinson's Integral Representation

We now turn to the problem of uniqueness. If Ω is bounded by finitely many analytic curves, there is a unique Garabedian function ψ, because the Szego kernel is unique. Since $f(\zeta)\psi(\zeta)d\zeta/ds = |\psi(\zeta)|$, and since ψ has no zeros on $\partial\Omega$, the Ahlfors function $f(z)$ is also unique, because it is determined by its boundary values. However there is another elementary argument which yields the uniqueness of the Ahlfors function for any compact set E.

Theorem 5.1: For any compact set E there is a unique function $f \in A(E,1)$ such that $f'(\infty) = \gamma(E)$.

Proof: Let E_n be compact sets decreasing to E, such that ∂E_n consists of finitely many pairwise disjoint analytic curves. Let $f_n(z) = \frac{\gamma(E_n)}{z} + \cdots$ be any Ahlfors function for E_n and let $\psi_n(z) = \frac{1}{2\pi i} + \cdots$ be a Garabedian function for E_n. Then for $g \in A(E_n,1)$ and $z \in \Omega_n = \mathbb{C}\setminus E_n$, we have by Cauchy's theorem

$$\int_{\partial\Omega_n} \frac{g(\zeta) - g(z)}{z - \zeta} \psi_n(\zeta)d\zeta = g(z) ,$$

so that

$$g(z) = \left(\int \frac{\psi_n(\zeta)}{\zeta - z} - 1\right)^{-1} \int \frac{g(\zeta)\psi_n(\zeta)}{\zeta - z} d\zeta .$$

Choose (n_j) so that the sequence $\{f_{n_j}\}$ converges pointwise to a function f in A(E,1), the sequence $\{\psi_{n_j}(\zeta)d\zeta\}$ converges weak star to a measure σ on E, and the sequence $\tau_{n_j} = \{f_{n_j}(\zeta)\psi_{n_j}(\zeta)d\zeta\} = \{|\psi_n(\zeta)||d\zeta|\}$ converges weak star to a positive measure τ on E. Then

$$f(z) = \lim_{j\to\infty}\left(\int \frac{\psi_{n_j}(\zeta)d\zeta}{\zeta - z} - 1\right)^{-1} \cdot \int \frac{f_{n_j}(\zeta)\psi_{n_j}(\zeta)}{\zeta - z}\, d\zeta$$

$$= \left(\int_E \frac{d\sigma(\zeta)}{\zeta - z} - 1\right)^{-1} \cdot \int_E \frac{d\tau(\zeta)}{\zeta - z} .$$

Let g be any extremal function for E: $g \in A(E,1)$, $g'(\infty) = \gamma(E)$. Then

$$g(z) = \left(\int_E \frac{d\sigma(\zeta)}{\zeta - z} - 1\right)^{-1} \lim_j \int \frac{g(\zeta)\psi_{n_j}(\zeta)}{\zeta - z}\, d\zeta .$$

We claim that the measures $\lambda_j = g(\zeta)\psi_{n_j}(\zeta)d\zeta$ converge weak star to τ. That clearly implies that $g = f$. Since

$$|g(\zeta)\psi_{n_j}(\zeta)d\zeta| \leq |\psi_{n_j}(\zeta)||d\zeta|$$

we have $|\lambda_j| \leq \tau_{n_j}$, while on the other hand

$$\int d\lambda_j = \int g(\zeta)\psi_{n_j}(\zeta)d\zeta = \gamma(E) = \lim_j \int d\tau_{n_j} = \int d\tau .$$

So if λ is a weak star limit of $\{\lambda_j\}$, then $|\lambda| \leq \tau$ and $\int d\lambda = \int d\tau$ so that $\lambda = \tau$. This means $\{\lambda_j\}$ converges to τ.

The above proof provides a representation of any $g \in A(E,1)$ as a fixed finction times a Cauchy transform. Let μ_g be a weak star limit of the sequence $g(\zeta)\psi_{n_j}(\zeta)d\zeta$. Then $\|\mu_g\| \leq \gamma(E)$ and

$$g(z) = \left(\int_E \frac{d\sigma(\zeta)}{\zeta - z} - 1\right)^{-1} \cdot \int \frac{d\mu_g(\zeta)}{\zeta - z}.$$

By Cauchy's theorem

$$\int \frac{d\sigma(\zeta)}{\zeta - z} = \lim_j \int \frac{\psi_{n_j}(\zeta)d\zeta}{\zeta - z}$$

$$= 1 - \lim_j 2\pi i\psi_{n_j}(z).$$

If we write $\psi(z) = \lim_j \psi_{n_j}(z)$, then the representation has the simpler form

$$g(z) = \frac{-1}{2\pi i\psi(z)} \int_E \frac{d\mu_g(\zeta)}{\zeta - z}.$$

More about this representation can be found in Havinson's paper [45]. Notice that if $\psi(z)$ is bounded below on Ω then every function in $A(E,1)$ is a Cauchy transform. An example in Chapter IV will show this is not always true.

There is another proof of uniqueness, due to Stephen Fisher [26], [28], which should be mentioned. The set $A(E,1)$ is convex and the functional $f \to f'(\infty)$ is linear. So there is a unique Ahlfors function if and only if each extremal function is an extreme point of this convex set. Fisher obtained uniqueness by showing any extremal function is an extreme point.

Problem 5.2: If E_n are compact sets decreasing to E, we know the Ahlfors functions for E_n converge to the Ahlfors function for E. If each ∂E_n consists of finitely many analytic Jordan curves, do the

Garabedian functions for E_n converge? If so, the limit is a natural definition of a Garabedian function for E. See Havinson's paper [45] for a discussion of this problem. It may help to observe that his measure μ^* is unique because of the Walsh-Lebesgue theorem [28].

Problem 5.3: Show there are constants C_1 and C_2 such that if E is a compact set and ∂E consists of finitely many pairwise disjoint analytic Jordan curves there is an "approximate Garabedian function"

$$\varphi(z) = \frac{1}{2\pi i} + \frac{b_1}{z} + \cdots$$

in $H^\infty(\Omega)$ such that $\|\varphi\|_\infty \leq C_1$ and

$$\int_{\partial E} |\varphi(\zeta)||d\zeta| \leq C_2\gamma(E) .$$

Quite likely, this is false.

Exercise 5.4: Show that a positive solution of Problem 5.3 would imply there is an absolute constant C such that

$$\gamma(E_1 \cup E_2) \leq C(\gamma(E_1) + \gamma(E_2))$$

for E_1 and E_2 compact.

Exercise 5.5: Assume the assertion in the above problem is true. Show there are constants C_3 and C_4 such that for any compact set E there is a measure μ on E such that

$$\hat{\mu}(z) = \int_E \frac{d\mu(\zeta)}{\zeta - z} \in A(E, C_3) .$$

$$\int d\mu = \gamma(E)$$

$$\|\mu\| \leq C_4 \gamma(E) .$$

§6. A Theorem of Pommerenke

A consequence of Theorem 4.1 is the following special result on semi-additivity of analytic capacity.

Theorem 6.1 (Pommerenke [66]): Let $E_1, E_2, \ldots, E_n$ be pairwise disjoint compact connected sets and let $K = E_1 + E_2 + \cdots + E_n$ be their Minkowski sum. Then

$$\gamma\left(\bigcup_{j=1}^{n} E_j\right) \leq \gamma(K) .$$

Proof: Let Ω be the complement of $E = \bigcup_{j=1}^{\infty} E_j$ and let $f \in A(E,1)$ satisfy $f'(\infty) = \gamma(E)$. Conformally mapping Ω to a domain with analytic boundary we see by 4.1 that

(i) f is an n to 1 covering map of $\{|w| < 1\}$.

(ii) f is proper: if $\Omega \ni z_j \to z \in E$, then $|f(z_j)| \to 1$.

Choose ρ, $0 < \rho < 1$ such that $L = f^{-1}(|w| = \rho)$ consists of n analytic curves and such that there are inverses $\varphi_1, \ldots, \varphi_n$ of f mapping $\rho < |w| < 1$ onto the n annular components of $\{z : \rho < |f(z)| < 1\}$. Now by Cauchy's theorem

$$\frac{1}{2\pi i} \int_L z(f(z))^k f'(z) dz = -\gamma(E)\delta_{k,0}$$

for $k \geq 0$. Therefore

$$\sum_{j=1}^{n} \frac{1}{2\pi i} \int_{|w|=\rho} \varphi_j(w) w^k dw = \gamma(E)\delta_{k,0} ,$$

and $\Phi(w) = \Sigma_{j=1}^{n} \varphi_j(w)$ extends to be meromorphic on $|w| < 1$ with a simple pole at 0:

$$\Phi(w) = \frac{\gamma(E)}{w} + a_0 + a_1 w + \cdots .$$

Let $T = \{\lim \Phi(w_n) : |w_n| \to 1\}$. Then T is a compact subset of K. Let U be the component of $\{|w| < 1\} \setminus \Phi^{-1}(T)$ containing the origin. Then Φ is a proper map of U onto $\Phi(U)$ and Φ has a simple pole at 0. Hence, by the argument principle Φ is univalent. Also $\partial\Phi(U) = T \subset K$, and Φ^{-1} is in $A(T,1)$ with derivative $\gamma(E)$ at ∞. Therefore, $\gamma(E) \leq \gamma(T) \leq \gamma(K)$.

An application is the explicit computation of $\gamma(E)$ for a linear set E.

<u>Theorem 6.2</u>: Let E be a subset of the real axis. Then

$$\gamma(E) = \ell(E)/4$$

where $\ell(E)$ is the inner arc length of E.

<u>Proof</u>: For E an interval this was proved in §1. First assume E is a union of pairwise disjoint closed intervals $\{E_j\}_{j=1}^{n}$. Then $K = E_1 + E_2 + \cdots + E_n$ is a closed interval of length $\ell(E)$. By Pommerenke's theorem

$$\gamma(E) \leq \gamma(K) = \ell(E)/4 .$$

Approximating from above yields the same inequality for all compact sets E; then approximating from below yields it for all E.

Conversely, let E be compact and write

$$f(z) = \frac{1}{2} \int_E \frac{dt}{t - z} .$$

Then $f(\infty) = 0$, $f'(\infty) = -\ell(E)/2$, and

$$|\operatorname{Im} f(z)| = \frac{|y|}{2} \int_E \frac{dt}{(t - x)^2 + y^2} , \quad z = x + iy$$

so that

$$|\operatorname{Im} f(z)| \leq \frac{1}{2} \int_{-\infty}^{\infty} \frac{du}{1 + u^2} = \frac{\pi}{2} .$$

Then $\operatorname{Re} e^{f(z)} > 0$ and

$$F(z) = \frac{1 - e^{f(z)}}{1 + e^{f(z)}} = \frac{\ell(E)}{4z} + \frac{a_2}{z^2} + \cdots$$

satisfies $|F(z)| \leq 1$. Hence $\gamma(E) \geq \ell(E)/4$, and this proves the theorem.

Notice that when E is compact the function $F(z)$ constructed above is the Ahlfors function for E.

Exercise 6.3: Let $E = \bigcup_{j=1}^{n} E_j$, where the E_j are pairwise disjoint closed intervals on the real axis. Prove that the Ahlfors function for E has exactly one zero in each of the bounded open intervals between the E_j.

Exercise 6.4: Let K be a compact subset of the real axis and let

$f \in A(K,1)$. Let $h(x) = \frac{1}{2\pi i} \lim_{y \downarrow 0} (f(x + iy) - f(x - iy))$. By Fatou's theorem $h(x)$ exists almost everywhere on K and is bounded. Prove

$$f(z) = \int_K \frac{h(x)dx}{z - x} ,$$

and show that $h(x) \geq 0$ when f is the Ahlfors function for K.

Exercise 6.5: Let E be a finite union of pairwise disjoint discs $\Delta(a_j, r_j)$ with centers a_j on the real axis and radii r_j. Show

$$\frac{1}{2} \sum r_j < \gamma(E) \leq \sum r_j .$$

Exercise 6.6: Derive Painlevé's theorem from Pommerenke's theorem.

§7. Analytic Capacity and Arc Length

We shall see in Chapter IV that the converse of Painleve's theorem does not hold. That is, $\gamma(E)$ can vanish for a set E of positive one dimensional Hausdorff measure. However when E lies on a sufficiently smooth curve $\gamma(E)$ can only vanish when E has zero length. The simplest result of this nature is Theorem 6.2 above and the sharpest result here is due to Ivanov [52]. Its intricate proof below is not used later and may be skipped.

Let Γ be a C^1 curve parametrized by arc length, $\Gamma = \{\zeta(s) : 0 \leq s \leq 1\}$. Then we can write $\zeta'(s) = e^{i\varphi(s)}$ where $\varphi(s)$ is real, and because our problem is a local one we can assume $|\varphi(s)| \leq 1/2$.

Theorem 7.1: (Ivanov). Let Γ be a C^1 curve such that for any s_0

$$(7.1) \qquad \int_0^1 \frac{|\varphi(s) - \varphi(s_0)|^2}{|s - s_0|} ds \le M$$

where M depends only on Γ. Then there is a constant $c_1(M)$ such that if E is a subset of Γ with inner arch length $\ell(E)$,

$$\gamma(E) \geq c_1(M)\ell(E) .$$

It will be clear from the proof that $c_1(M) = (aM+b)^{-1}$ for some a and b. Condition (7.1) holds whenever the modulus of continuity $\omega_\varphi(\delta)$ of φ satisfies

$$\int_0 \frac{\omega_\varphi(\delta)^2 d\delta}{\delta} < \infty .$$

This holds for any <u>Lyaponov curve</u>, which by definition is a curve for which $\omega_\varphi(\delta)$ satisfies a Hölder condition $\omega_\varphi(\delta) = O(\delta^\beta)$, $\beta > 0$. It also holds for the <u>hypo-Lyaponov</u> curves defined in [18] by

$$\int_0 \frac{\omega_\varphi(\delta) d\delta}{\delta} < \infty .$$

<u>Corollary 7.2</u>: If Γ satisfies the conditions of Theorem 7.1 there is $C(\Gamma)$ such that

$$\gamma(E_1 \cup E_2) \leq C(\Gamma)(\gamma(E_1) + \gamma(E_2))$$

whenever E_1 and E_2 are Borel subsets of Γ.

<u>Lemma 7.3</u>: Let Γ be a C^1 curve $\{\zeta(s) : 0 \le s \le 1\}$ such that $\zeta'(s) = e^{i\varphi(s)}$, $|\varphi(s)| \le 1/2$. Let $|\theta(s)| \le 1$ and let $\mathrm{dist}(z,\Gamma) = \varepsilon = |z - \zeta(s_0)|$. Then

$$\left|\int_\Gamma \frac{\theta(s)ds}{\zeta(s) - z} - \int_{\Gamma\cap\{|s-s_0|>\varepsilon\}} \frac{\theta(s)ds}{\zeta(s) - \zeta(s_0)}\right| \leq 5 .$$

Proof: We must estimate

$$\int_{|s-s_0|<\varepsilon} \frac{\theta(s)ds}{\zeta(s) - z} + (z - \zeta(s_0)) \int_{|s-s_0|>\varepsilon} \frac{\theta(s)ds}{(\zeta(s) - \zeta(s_0))(s - s_0)} = I_1 + I_2 .$$

Clearly $|I_1| \leq 2$, while

$$|I_2| \leq \varepsilon \int_{|s-s_0|>\varepsilon} \frac{ds}{|\zeta(s) - \zeta(s_0)||s - s_0|} .$$

Since $|\zeta(s) - \zeta(s_0)| = |\int_{s_0}^{s} e^{i\varphi(t)}dt| \geq \cos\frac{1}{2} \cdot |s - s_0|$, we have

$$|I_2| \leq \frac{2\varepsilon}{\cos\frac{1}{2}} \int_\varepsilon^1 \frac{dt}{t^2} \leq \frac{2}{\cos\frac{1}{2}} \leq 3 .$$

Proof of 7.1: We of course assume E is compact and $\ell(E) > 0$. By Exercise 6.4 there is a non-negative function $\theta(s)$ on $\zeta^{-1}(E)$ such that $\theta \leq 2$,

$$\left|\int \frac{\theta(s)ds}{s - z}\right| \leq 1 \quad \text{if} \quad z \notin \zeta^{-1}(E) ,$$

and $\int \theta(s)ds = -\ell(E)/4$. Write

$$f(z) = \int \frac{\theta(s)ds}{\zeta(s) - z} .$$

Then f is analytic off E and $f'(\infty) = -\ell(E)/4$. We claim, for $\zeta(s_0) \in E$

$$(7.2)\qquad \text{Real}\left(e^{i\varphi(s_0)}\int_{|s-s_0|>\varepsilon}\frac{\Theta(s)ds}{\zeta(s)-\zeta(s_0)}\right)\le C_2M+12\ .$$

Assuming this for the moment, we have by the lemma

$$\text{Real } e^{i\varphi(s_0)} f(z)\le C_2M+17=\lambda\ ,$$

so that

$$\left|f(z)-\frac{2\lambda}{\cos\frac{1}{2}}\right| = \left|e^{i\varphi(s_0)}f(z)-\frac{2\lambda}{\cos\frac{1}{2}}e^{i\varphi(s_0)}\right|$$

$$\ge \frac{2\lambda\cos\varphi(s_0)}{\cos\frac{1}{2}}-\lambda\ge\lambda\ .$$

Writing $\beta = 2\lambda(\cos\frac{1}{2})^{-1}$ we have $|f(z)-\beta|>\lambda$ so that

$$F(z)=\frac{\lambda f(z)}{\beta f(z)-(\beta^2-\lambda^2)}$$

is in $A(E,1)$ and $|F'(\infty)|\ge\dfrac{\lambda\ell(E)}{4(\beta^2-\lambda^2)}=c_1(M)\ell(E)$.

To verify (7.2) we show

$$A=\text{Real}\left(e^{i\varphi(s_0)}\int_{|s-s_0|>\varepsilon}\frac{\Theta(s)ds}{\zeta(s)-\zeta(s_0)}-\int_{|s-s_0|>\varepsilon}\frac{\Theta(s)ds}{s-s_0}\right)\le C_2M$$

and apply the lemma. Introduce the notation

$$\zeta(s)-\zeta(s_0)=e^{i\varphi(s_0)}\int_{s_0}^{s}e^{i\varphi(t)-i\varphi(s_0)}dt=e^{i\varphi(s_0)}\rho(s_0,s)e^{i\alpha(s_0,s)}$$

where $\rho(s_0,s)\ge 0$ and $|\alpha(s_0,s)|\le 1$. Then

$$A=\int_{|s-s_0|>\varepsilon}\Theta(s)\,\frac{[(s-s_0)\cos\alpha(s_0,s)-\rho(s_0,s)]}{\rho(s_0,s)(s-s_0)}\,ds\ ,$$

and as $\rho(s_0,s) \geq |s - s_0|\cos\frac{1}{2}$, we have

$$A \leq \frac{2}{\cos\frac{1}{2}} \int_{|s-s_0|>\varepsilon} \frac{1}{|s - s_0|^2} [\cos\alpha(s_0,s)|s - s_0 - \rho\cos\alpha| + \rho\sin^2\alpha]ds \ .$$

But

$$\cos\alpha|s - s_0 - \rho\cos\alpha| \leq \left|\int_0^{s-s_0} 1 - \cos(\varphi(s_0 + t) - \varphi(s_0))dt\right|$$

$$\leq C_3 \int_0^{s-s_0} |\varphi(s_0 + t) - \varphi(s_0)|^2 dt \ ,$$

and

$$\rho\sin^2\alpha = \frac{1}{\rho}\left(\operatorname{Im}\int_0^{s-s_0} e^{i\varphi(s_0+t)-i\varphi(s_0)}dt\right)^2$$

$$\leq \frac{1}{\rho}\left(\int_0^{s-s_0} |\varphi(s_0 + t) - \varphi(s_0)|dt\right)^2 \leq C_4 \int_0^{s-s_0} |\varphi(s_0 + t) - \varphi(s_0)|^2 dt \ .$$

Thus

$$A \leq C_2 \int_\varepsilon^1 \frac{1}{u^2} \int_0^u |\varphi(s_0 + t) - \varphi(s_0)|^2 dt du \ ,$$

so that by Fubini's theorem

$$A \leq C_2 \int_0^1 \frac{|\varphi(s_0 + t) - \varphi(s_0)|^2}{t} dt \leq C_2 M \ .$$

<u>Problem 7.4</u>: If Γ is a rectifiable curve, or even a C^1 curve, and E is a compact subset of Γ with positive length, show $\gamma(E) > 0$.

This was incorrectly proved by Denjoy [21], and the assertion is now called the Denjoy conjecture. It would be a consequence of the semi-additivity [18].

CHAPTER II. THE CAUCHY TRANSFORM

§1. Basic Properties

Let μ be a finite, compactly supported measure. The Cauchy transform of μ is the function

$$\hat{\mu}(z) = \int \frac{d\mu(\zeta)}{\zeta - z}, \tag{1.1}$$

defined for all z for which the Newtonian potential

$$U_{|\mu|}(z) = \int \frac{d|\mu|(\zeta)}{|\zeta - z|} \tag{1.2}$$

is finite.

Theorem 1.1: The Cauchy transform $\hat{\mu}$ is absolutely convergent for almost all z, and satisfies

(i) $\hat{\mu} \in L^p_{loc}$ for $1 \le p < 2$

(ii) $\hat{\mu}$ is analytic off the closed support S_μ of μ.

(iii) $\hat{\mu}(\infty) = 0$

(iv) $\hat{\mu}'(\infty) = \lim_{z\to\infty} z\mu(z) = -\int d\mu$.

Proof: Since for $1 \le p < 2$, $1/|z| \in L_p^{loc}$, Fubini's theorem shows that $U_{|\mu|}$ converges almost everywhere and that $\hat{\mu}$ (even $U_{|\mu|}$) is in L^p_{loc}. Properties (ii), (iii) and (iv) are verified by routine calculations.

In Chapter III it will be proved that the set where $U_{|\mu|}$ diverges is much smaller than a set of area zero. It has Newtonian capacity zero and hence Hausdorff dimension at most one.

It is often convenient to regard $\hat{\mu}$ as the solution of an inhomogeneous Cauchy-Riemann equation.

<u>Theorem 1.2</u>: If μ is a finite, compactly supported measure, then

$$\frac{\partial \hat{\mu}}{\partial \bar{z}} = -\pi\mu \tag{1.3}$$

in the sense of distributions.

<u>Proof</u>: By definition, formula (1.3) means

$$\int g(\zeta)d\mu(\zeta) = \frac{1}{\pi} \iint \frac{\partial g}{\partial \bar{z}} \hat{\mu}(z)dxdy$$

for all $g \in C_0^\infty$. Since by Green's theorem

$$g(\zeta) = \frac{1}{\pi} \iint \frac{\partial g}{\partial \bar{z}} \frac{1}{\zeta - z} dxdy ,$$

we get (1.3) via Fubini's theorem.

<u>Corollary 1.3</u>: If $\hat{\mu}$ is almost everywhere equal to a function analytic on an open set V, then $|\mu|(V) = 0$. If $\hat{\mu} = 0$ a.e., then $\mu = 0$.

<u>Proof</u>: This follows from (1.3) and the density of C_0^∞ in the compactly supported continuous functions.

Condition (1.3) and the obviously necessary conditions $\hat{\mu} \in L^1_{loc}$, $\hat{\mu}(\infty) = 0$ actually determine the Cauchy transform $\hat{\mu}$ of μ. This fact is a variant of Weyl's lemma.

<u>Theorem 1.4</u>: Let $f \in L^1_{loc}$ satisfy $\lim_{z\to\infty} f(z) = 0$ and let μ be a

compactly supported measure such that $\frac{\partial f}{\partial \bar{z}} = -\pi\mu$ as distributions. Then $f(z) = \hat{\mu}(z)$ almost everywhere.

Proof: Replacing f by $f - \hat{\mu}$ we assume $f \in L^1_{loc}$, $f(\infty) = 0$ and $\frac{\partial f}{\partial \bar{z}} = 0$ weakly, and we must show $f = 0$ a.e. Let χ_ρ be a C_0^∞ approximate identity: $\chi_\rho \geq 0$, $\int \chi_\rho \, dxdy = 1$, $\chi_\rho = 0$ off $\Delta(0,\rho) = \{z : |z| < \rho\}$. Then the convolution

$$f_\rho(z) = f * \chi_\rho(z) = \iint f(z - \zeta)\chi_\rho(\zeta)d\xi d\eta \qquad \zeta = \xi + i\eta$$

is in C^∞, and f_ρ converges to f in $L^1(K)$ for any compact set K. Moreover

$$\frac{\partial f_\rho}{\partial \bar{z}} = \frac{\partial f}{\partial \bar{z}} * \chi_\rho = 0$$

by Fubini's theorem and the definition of weak derivative (see [51, p. 14]). Therefore f_ρ is entire and vanishes at ∞, so that $f_\rho = 0$, and hence $f = 0$ a.e.

§2. A Characterization of Cauchy Transforms

Let f be a locally integrable function. Assume that f is analytic off a compact set E and that $f(\infty) = 0$. In this section we determine when there is a measure μ on E such that $f(z) = \hat{\mu}(z)$ area almost everywhere. We begin by recapturing μ from $\hat{\mu}$ using Cauchy's theorem.

Lemma 2.1: Let Γ be a rectifiable Jordan curve bounding a domain D.

If μ is a finite, compactly supported measure such that

$$(2.1) \qquad \int_\Gamma U_{|\mu|}(z)ds < \infty$$

then

$$(2.2) \qquad \frac{-1}{2\pi i} \int_\Gamma \hat{\mu}(z)dz = \mu(D) = \mu(\overline{D}) .$$

Proof: Since $\int_\Gamma \frac{ds}{|\zeta - z|} = \infty$ for $\zeta \in \Gamma$, (2.1) implies $|\mu|(\Gamma) = 0$. Fubini's theorem then yields (2.2).

Though trivial, Lemma 2.1 is very useful because, being locally integrable, $U_{|\mu|}$ is integrable over almost all rectangles. This fact is the key to the theorem characterizing Cauchy transforms.

Call a collection $\mathfrak{R}$ of open rectangles, each with its sides parallel to the axes, an admissible grid if the set of pairs (z,w) where z and w respectively are the lower left and upper right vertices of a rectangle in $\mathfrak{R}$ has full measure in $\{(z,w) \in \mathbb{C} \times \mathbb{C} : \operatorname{Re} w > \operatorname{Re} z, \operatorname{Im} w > \operatorname{Im} z\}$. Clearly a countable intersection of admissible grids is again an admissible grid. If f is locally integrable, there is an admissible grid $\mathfrak{R}$ such that

$$\int_{\partial R} |f(z)|ds < \infty$$

for all $R \in \mathfrak{R}$, because $|f(z)|$ is locally integrable over almost all horizontal and vertical lines.

Lemma 2.2: (Royden [70]). Let D be a domain, $f \in L^1_{loc}(D)$ and assume

$$\int_{\partial R} |F(z)|ds < \infty, \quad \int_{\partial R} F(z)dz = 0$$

for all rectangles R whose closures lie in D and which belong to some admissible grid $\mathcal{R}$. Then F is almost everywhere equal to a function analytic on D. In particular, if $D = \mathbb{C}$ and F vanishes at ∞, then $F = 0$ a.e.

<u>Proof</u>: Let $\{\chi_\rho\}$ be the approximate identity introduced in the proof of Theorem 1.4. Then $F * \chi_\rho$ is in $C^\infty(D)$, and for $R \in \mathcal{R}$ with $\overline{R} \subset D$ and ρ small, we have by Fubini's theorem

$$\int_{\partial R} (F * \chi_\rho)(z)dz = \int_{\partial R}\iint F(z - \zeta)\chi_\rho(\zeta)d\xi d\eta dz$$

$$= \iint \chi_\rho(\zeta) \int_{\partial(R-\zeta)} F(w)dwd\xi d\eta = 0$$

because almost every translate $R - \zeta$ of R is in $\mathcal{R}$. Morera's theorem now implies $F * \chi_\rho$ is analytic on D. As ρ tends to 0, $F * \chi_\rho$ converges almost everywhere to F and converges uniformly on compact sets to an analytic function g, so that $F = g$ almost everywhere.

<u>Theorem 2.3</u>: Let $f \in L^1_{loc}$ be analytic off a compact set E and satisfy $f(\infty) = 0$. Let μ be a measure on E. The following are equivalent

(i) $f(z) = \hat{\mu}(z)$ a.e.

(ii) There is an admissible grid $\mathcal{R}$ such that

$$\int_{\partial R} |f(z)|ds < \infty, \quad \frac{-1}{2\pi i}\int_{\partial R} f(z)dz = \mu(R)$$

for all $R \in \mathcal{R}$.

Proof: Assume (i) holds. Then the set of rectangles R for which $\int_{\partial R} U_{|\mu|}(z)ds < \infty$ forms an admissible grid and for each such R

$$\frac{-1}{2\pi i} \int_{\partial R} f(z)dz = \mu(R)$$

by Lemma 2.1.

Now assume (ii) holds. Let $F(z) = f(z) - \hat{\mu}(z)$. Then $F \in L^1_{loc}$ and F is analytic off E and vanishes at ∞. Moreover, there is an admissible grid $\mathcal{R}$ such that

$$\int_{\partial R} |F(z)|ds < \infty, \quad \int_{\partial R} F(z)dz = 0$$

for all $R \in \mathcal{R}$. But this implies $F = 0$ a.e., by Lemma 2.2.

Corollary 2.4: Let E be a closed subset of a domain D and let f be locally integrable and analytic on $D \setminus E$. Assume $\int_\gamma f(z)dz = 0$ for all curves γ in $D \setminus E$. If the projections of E onto the coordinate axes have zero length, then f extends to be analytic on D.

Proof: This is actually a consequence of 2.2. The set of rectangles in D whose boundaries miss E forms an admissible grid and so 2.2 shows that f extends analytically to D.

This result is extended somewhat in the next section. The example in §3 of IV shows that the metric condition put on the set E in 2.4 cannot be weakened.

We now describe these $f \in L^1_{loc}$ which are almost everywhere

Cauchy transforms. Call a measurable function f <u>of bounded variation</u> if there is an admissible grid $\mathcal{R}$ such that

(2.3) $$\int_{\partial R} |f(z)|\,ds < \infty \quad \text{for all} \quad R \in \mathcal{R}$$

and there is a number M such that whenever $R_1, \ldots, R_m \in \mathcal{R}$ are pairwise disjoint

(2.4) $$\sum_{j=1}^{m} \left| \frac{1}{2\pi i} \int_{\partial R_j} f(z)\,dz \right| \leq M \, .$$

Of course the number M may depend on the grid $\mathcal{R}$. The infimum of the numbers M which arise as we vary the grid $\mathcal{R}$ will be called the <u>total variation</u>, $V(f)$, of f. The total variation of the restriction of f to a open subset U will be denoted by $V(f,U)$ and is defined by (2.4) with the constraint that only rectangles contained in U are used. The next theorem is essentially proved in [76], but it may have been noticed even earlier.

<u>Theorem 2.5</u>: Assume that f is locally integrable, analytic off a compact set E and $f(\infty) = 0$. Then there is a measure μ on E such that

$$f(z) = \hat{\mu}(z) \quad \text{a.e.}$$

if and only if f is of bounded variation. When this is the case the total variation $V(f)$ is attained for any grid $\mathcal{R}$ for which

$$\int_{\partial R} U_{|\mu|}\,ds < \infty$$

for all $R \in \mathcal{R}$. Moreover

$$V(f;\, U) = |\mu|(U)$$

for any open set U.

Theorem 2.5 is analogous to a familiar result on functions of one variable: the function f on the real line is of bounded variation if and only if there is a finite measure ν such that

$$f(b) - f(a) = \int_a^b d\nu(x) = \nu([a,b)) \ .$$

In this case $V_a^b(f) = |\nu|([a,b))$ and $df/dx = \nu$. We have merely replaced $f(b) - f(a)$ by

$$\frac{-1}{2\pi i} \int_{\partial R} f dz \ ,$$

and d/dx by $-1/\pi\, \partial/\partial\bar{z}$. A simple, albeit slightly sophisticated, proof of the classical fact consists of convolving with an approximate identity and applying the fundamental theorem of calculus. Similarly, we will use Green's theorem and the functions χ_ρ.

<u>Proof of 2.5</u>: If $f(z) = \hat{\mu}(z)$ a.e., the conclusions follow from Lemma 2.1. So our task is to prove that every function of bounded variation is a Cauchy transform. Now assume f is of bounded variation and let $\mathcal{R}$ be an admissible grid for which (2.3) and (2.4) hold for f. Discard

from $\mathfrak{R}$ any rectangle R such that

$$\lim_{\delta\to 0} V(f,U_\delta) > 0$$

where U_δ is the δ-neighborhood of ∂R. Since f is of bounded variation there is, for any $\varepsilon > 0$, a finite set of horozontal and vertical lines such that if R is a rectangle for which $\lim_{\delta\to 0} V(f,U_\delta) > \varepsilon$, then ∂R is covered by this collection of lines. Hence the new smaller grid $\mathfrak{R}$ arises by deleting certain rectangles whose boundaries lie on at most a countable number of lines. This new grid is still an admissible grid for which (2.3) and (2.4) hold, and it has the additional property

$$\lim_{k\to\infty} \int_{\partial R_k} f(z)dz = \int_{\partial R} f(z)dz \tag{2.5}$$

whenever (the vertices of) the rectangles R_k converge to (the vertices of) $R \in \mathfrak{R}$.

Let χ_ρ be the approximate identity introduced above, and write $f_\rho = f * \chi_\rho \in C^\infty$. By Green's theorem, f_ρ is the Cauchy transform of

$$\mu_\rho = \frac{-1}{\pi} \frac{\partial f_\rho}{\partial \bar{z}} dxdy .$$

For $R \in \mathfrak{R}$ we have

$$\mu_\rho(R) = \frac{-1}{2\pi i} \int_{\partial R} f_\rho(z)dz$$

$$= \frac{-1}{2\pi i} \int_{\partial R} \iint f(z - \zeta)\chi_\rho(\zeta)d\xi d\eta\, dz$$

$$= -\iint \chi(\zeta) \frac{1}{2\pi i} \int_{\partial(R-\zeta)} f(w)dw d\xi d\eta \ .$$

This means that for any $R \in \mathcal{R}$

$$\lim_{\rho\to 0} \mu_\rho(R) = \frac{-1}{2\pi i}\int_{\partial R} f(z)dz \tag{2.6}$$

by (2.5). It also means that if $R_1,\dots,R_n \in \mathcal{R}$, then

$$\sum_j |\mu_\rho(R_j)| \le \iint \chi_\rho(\zeta) \sum_j \left|\frac{1}{2\pi i} \int_{\partial(R_j-\zeta)} f(w)dw\right| d\xi d\eta \ . \tag{2.7}$$

Consequently $\|\mu_\rho\| \le V(f)$. Let μ be a weak star limit of the $\{\mu_\rho\}$. Let $R \in \mathcal{R}$, $\varepsilon > 0$ and let δ be so small that $V(f,U_\delta) < \varepsilon/2$, where U_δ is the δ-neighborhood of ∂R. Then choose $g \in C_o(R)$ such that $0 \le g \le 1$, $g = 1$ on $R\backslash U_{\delta/2}$ and

$$\left|\int g d\mu - \mu(R)\right| < \varepsilon \ .$$

By (2.7) we have $|\mu_\rho|(U_{\delta/2}) = V(f_\rho,U_{\delta/2}) < \varepsilon/2$ if $\rho < \delta/2$. Therefore, if $\rho < \delta/2$

$$\left|\mu_\rho(R) - \int g d\mu_\rho\right| \le 2|\mu_\rho|(U_{\delta/2}) < \varepsilon$$

and as $\int g d\mu_\rho \to \int g d\mu$, we have

$$\mu_\rho(R) \to \mu(R) \qquad R \in \mathcal{R} \ .$$

With (2.6) this shows $f(z) = \hat{\mu}(z)$ a.e. by Theorem 2.3. By Corollary 1.3, μ is supported on E.

There is an alternate proof, which we merely outline. Simply define the function ν on $\mathcal{R}$ by

$$\nu(R) = \frac{-1}{2\pi i} \int_{\partial R} f(z)dz .$$

Then (2.4) implies that ν has a countably additive, bounded extension to the Borel sets. This can be seen via the Riesz representation theorem, or directly using the usual exhaustion arguments.

When f is continuous and analytic off a compact set E we can estimate its variation using less fine systems of rectangles. For example, let $\mathcal{G}$ denote the grid of closed squares of side 2^{-n}, $n \in Z$, with vertices at the lattice points $(p + iq)2^{-n}$, $p,q \in Z$. Set

$$V_{\mathcal{G}}(f) = \sup \sum_{j=1}^{m} \left| \frac{1}{2\pi i} \int_{\partial S_j} f(z)dz \right| ,$$

where the supremum is taken over all finite coverings $\{S_1, \ldots, S_m\}$ of E by squares from $\mathcal{G}$.

<u>Corollary 2.6</u>: Let f be continuous on the Riemann sphere and analytic off a compact set E. Assume $f(\infty) = 0$. Then there is a measure μ on E such that $f = \hat{\mu}$ a.e. if and only if $V_{\mathcal{G}}(f) < \infty$. When this is the case $\|\mu\| = V_{\mathcal{G}}(f)$.

<u>Proof</u>: From the continuity it follows that the (possibly infinite) variation of f is attained over the grid $\mathcal{R}$ of all rectangles, and also that $V(f) = V_{\mathcal{G}}(f)$.

Exercise 2.7: Let $f \in C(E,1)$ satisfy a Lipschitz condition $|f(z) - f(w)| \leq K|z - w|$. Then f is the Cauchy transform of a measure $h\,dxdy$, $h \in L^{\infty}$. If the set F has area zero, then f is in the uniform closure of $C(E\setminus F,1)$.

Exercise 2.8: Let μ be a measure on a compact set E such that $\hat{\mu} \in C(E,1)$. Prove that $|\mu|(L) = 0$ for every straight line L. Prove that $|\mu|(J) = 0$ if J is a rectifiable curve.

§3. Painlevé Length, Pompeiu Variation, and a Theorem of Havin

Throughout this section we fix a compact set E and a function f, supposed, as before, to be analytic on $S^2\setminus E$ and to vanish at ∞. We want to know when there is a measure μ on E such that $f(z) = \hat{\mu}(z)$ for all $z \notin E$. If E has zero area, Theorem 2.5 provides an answer, but to apply 2.5 when E has positive area we must determine when f has a measurable extension at bounded variation on $\mathbb{C}$. It is easier to attack the problem directly.

By a regular neighborhood of E we mean an open set $V \supset E$ such that ∂V consists of finitely many analytic Jordan curves surrounding E in the usual sense of contour integration. We say that the set E has finite Painlevé length if there is a number ℓ such that every open $U \supset E$ contains regular neighborhood V of E such that ∂V has length at most ℓ. The infimum of such numbers ℓ is called the Painlevé length of E. Our first theorm, due to Erohin and Havinson [45], is a slight generalization of Painlevé's theorem.

Theorem 3.1: Let E be a compact set with finite Painlevé length κ. If $f \in A(E,m)$ then there is a measure μ on E such that

$$\|\mu\| \le \frac{\kappa}{2\pi} m ,$$

$$\hat{\mu}(z) = f(z) \qquad z \notin E .$$

Consequently,

$$\gamma(E) \le \frac{\kappa}{2\pi} .$$

Proof: Let $\{V_n\}_{n=1}^{\infty}$ be a decreasing sequence of regular neighborhoods of E such that $\bigcap V_n = E$ and ∂V_n has total length at most $\kappa + 1/n$. For $z \notin V_n$, we have

$$f(z) = \frac{-1}{2\pi i} \int_{\partial V_n} \frac{f(\zeta)d\zeta}{\zeta - z} = \int_{\partial V_n} \frac{d\mu_n(\zeta)}{\zeta - z}$$

where $\|\mu_n\| \le m(\kappa + \varepsilon)/2\pi$. The μ_n converge weak star as measures on $\overline{V}_1$ to a measure μ on E such that $\|\mu\| \le m\kappa/2\pi$ and $\hat{\mu}(z) = f(z)$ for $z \notin E$.

Of course the measures μ_n on ∂V_n could be bounded even if E had infinite Painlevé length. Thus we have

Theorem 3.2: Let E be compact and let f be analytic on $S^2 \setminus E$ with $f(\infty) = 0$. Assume there is a decreasing sequence $\{V_n\}_{n=1}^{\infty}$ of regular neighborhoods of E such that $\bigcap_n \overline{V}_n = E$, and

$$\frac{1}{2\pi} \int_{\partial V_n} |f(\zeta)| ds \le M .$$

Then there is a measure μ on E with $\|\mu\| \le M$ such that $\hat{\mu}(z) = f(z)$ for all $z \notin E$.

There is a converse theorem of a similar nature. The Pompeiu variation of f is the supremum

$$V_E(f) = \sup \sum_{j=1}^{n} \left| \frac{1}{2\pi i} \int_{\Gamma_j} f(z)dz \right|$$

taken over all collections $\{\Gamma_j\}_{j=1}^{n}$ of curves such that $\bigcup \Gamma_j$ is the boundary of a regular neighborhood of E. If L is an open-closed subset of E, write $f = f_1 + f_2$ where f_1 is analytic on $S^2\setminus L$ and f_2 is analytic on a neighborhood of L and define

$$V_L(f) = V_L(f_1) .$$

Theorem 3.3: Let E be compact and let f be analytic on $S^2\setminus E$ with $f(\infty) = 0$. Assume there is a measure μ on E with $f(z) = \hat{\mu}(z)$ for $z \notin E$. Then

(a) For any open-closed subset of L of E,

$$|\mu(L)| \leq V_L(f) \tag{3.1}$$

$$V_E(f) = V_L(f) + V_{E\setminus L}(f) . \tag{3.2}$$

(b) If E is totally disconnected and $f \not\equiv 0$,

$$0 < V_E(f) < \infty \tag{3.3}$$

and for any open-closed subset L

$$V_L(f) = |\mu|(L) . \tag{3.4}$$

Proof: For part (a) write $\mu_1 = \mu|_L$. Then $f_1 = \hat{\mu}_1$ so that we may assume $L = E$. Then

$$|\mu(E)| = \left| \int \frac{1}{2\pi i} \int_{\partial V} \frac{dz}{\zeta - z} \, d\mu(\zeta) \right|$$

$$= \left| \sum_{j=1}^{n} \frac{1}{2\pi i} \int_{\Gamma_j} \hat{\mu}(z) dz \right| \leq V_E(f) ,$$

and (3.1) holds. Statement (3.2) is immediate from the definitions.

To prove (b) notice that when E is totally disconnected the supremum $V_L(f)$ is attained using only regular neighborhoods of E which have simply connected components. Then each curve Γ_j bounds a component V_j of V and

$$(3.5) \qquad \left| \frac{1}{2\pi i} \int_{\Gamma_j} f_1(z) dz \right| = |\mu(V_j)| .$$

Since for any $\varepsilon > 0$ there is a decomposition $L = \bigcup_{k=1}^{s} L_k$ of L into pairwise disjoint open-closed subsets such that

$$|\mu|(L) - \varepsilon \leq \sum |\mu(L_k)| \leq |\mu|(L) ,$$

we obtain (3.4) from (3.5), (3.1), and (3.2). Finally (3.3) follows from (3.4).

Corollary 3.4: Let E be a compact set such that the projections of E onto the coordinate axes have zero length. Let $f \in A(E,1)$ have finite Pompeiu variation. Then there is a unique measure μ on E such that $f(z) = \hat{\mu}(z)$ almost everywhere.

Proof: This follows from 3.3(b) and 2.4.

The converses of 3.2 and 3.3(b) do not hold (see Exercises 3.8 and 3.9 below). However, Havin's theorem [40] does provide a necessary and sufficient condition for a function to be a Cauchy transform. This theorem also has 3.1 and 3.3(b) as simple corollaries.

Suppose $f(z) = \hat{\mu}(z)$, $z \notin E$ where μ is a measure on E. Let V be a regular neighborhood of E and let g be analytic on $\overline{V}$. The integral

$$T_f(g) = \frac{-1}{2\pi i} \int_{\partial V} f(z)g(z)dz$$

is independent of V and satisfies $|T_f(g)| \leq \|g\|_E \|\mu\|$, where $\|g\|_E = \sup_{z\in E} |g(z)|$, because

$$T_f(g) = \frac{-1}{2\pi i} \int_{\partial V} \int_E \frac{d\mu(\zeta)}{\zeta - z} g(z)dz$$

$$= \int_E g(\zeta)d\mu(\zeta) \ .$$

Thus a necessary condition that $f = \hat{\mu}$ is that the linear functional T_f be bounded with respect to the norm $\|g\|_E$. Using the Hahn-Banach and Riesz representation theorems, we see that this condition is also sufficient.

Theorem 3.5 (Havin): Let E be a compact set and let f be analytic on $S^2 \setminus E$ and satisfy $f(\infty) = 0$. There is a measure μ on E such that $f(z) = \hat{\mu}(z)$ for $z \notin E$ if and only if there is a constant C_f such that

$$(3.6) \qquad |T_f(g)| \le c_f \|g\|_E$$

for all g analytic near E. When this is the case we may take $c_f = \|\mu\|$.

Proof: We have already shown that (3.6) is necessary. Now assume (3.6). Then the linear functional T_f has an extension to $C(E)$ of norm c_f and this extension is represented by a measure μ. Let $g(\zeta) = 1/(\zeta - z)$ where $z \notin E$. Let V be a regular neighborhood of E with $z \notin \overline{V}$. Then

$$T_f(g) = \frac{1}{2\pi i} \int_{\partial V} \frac{f(\zeta)d\zeta}{\zeta - z} = f(z)$$

and

$$T_f(g) = \int g(\zeta)d\mu(\zeta) = \hat{\mu}(z) .$$

Corollary 3.6: Let E be compact and let f be analytic on $S^2\setminus E$ with $f(\infty) = 0$. There is a measure μ on E such that $f(z) = \hat{\mu}(z)$ for all $z \notin E$ if and only if there is $M < \infty$ such that

$$\left|\sum_{k=1}^{n} \lambda_k f(a_k)\right| \le M \sup_{z\in E} \left|\sum \frac{\lambda_k}{z - a_k}\right|$$

whenever $a_1,\ldots,a_n \notin E$ and $\lambda_k \in \mathbb{C}$. In this case we have $\|\mu\| \le M$.

Proof: By Runge's theorem every function analytic near E is a uniform limit of functions of the form

$$g(z) = \sum_{k=1}^{n} \frac{\lambda_k}{z - a_k} , \quad a_k \notin E .$$

But

$$|T_f(g)| = \left|\sum \lambda_k f(a_k)\right| .$$

Exercise 3.7: Derive Theorems 3.2 and 3.3(b) from Havin's theorem.

Exercise 3.8: In Chapter IV there is an example of a totally disconnected compact set K and a function $f \in C(K,1)$ such that $f \not\equiv 0$ but $V_K(f) = 0$. Prove that there is a totally disconnected compact set E and $g \in A(E,1)$ such that

$$0 < V_E(g) < \infty$$

but g is not a Cauchy transform.

Exercise 3.9: Show the converse of 3.2 fails. One strategy is outlined below, but there could be a simpler construction. For any positive integer n construct a set $E_n = K_n \cup L_n$ where K_n is a totally disconnected compact subset of the disc $|z| < 1/n^2$ with $\text{area}(K_n) \geq 1/n^4$, and L_n is a compact totally disconnected subset of the annulus $A_n = \{1/2n \leq |z| \leq 1/n\}$ with $\gamma(L_n) > 0$, such that if V is a regular neighborhood of E_n sufficiently close to E_n then $A_n \cap \partial V$ has length exceeding n. Show there exists μ_n on E_n such that $\|\mu_n\| \leq 1/n$, $\hat{\mu}_n \in A(E_n,1)$, but for V a regular neighborhood sufficiently close to E_n,

$$\int_{\partial V} |\hat{\mu}(z)|\, ds > c$$

where c is a constant independent of n. Now show there exists a totally disconnected compact set E and a measure μ such that $S_\mu = E$,

$\hat{\mu} \in A(E,1)$ but

$$\lim_{n\to\infty} \int_{\partial V_n} |\hat{\mu}(z)|ds = \infty$$

for every sequence V_n of regular neighborhoods decreasing to E.

Exercise 3.10: Let $f = \Sigma_{n=0}^{\infty} a_n z^n$ be analytic on the unit disc. If μ is a measure on the unit circle such that $f(z) = \hat{\mu}(z)$, $|z| < 1$ show $|a_n|$ is bounded. Show the converse fails.

§4. Two Problems on Cauchy Transforms

Problem 4.1: If $\gamma(E) > 0$, is there a non-zero complex measure μ on E such that $\hat{\mu} \in A(E,1)$? Probably this is false. If it is true there is a constant K, independent of E, and a non-zero measure μ on E with $\hat{\mu} \in A(E,1)$ and $\|\mu\| \leq K|\int d\mu|$, although the proof of this is not trivial. A counterexample would give a negative solution to the problem on the "approximate Garabedian function" posed in Chapter I.

Problem 4.2: Describe the sets whose characteristic functions are Cauchy transforms almost everywhere.

A solution of this problem would yield information on the Gleason parts for the uniform algebra $R(K)$. A Swiss cheese is a compact set obtained by deleting from the closed unit disc $\bar{\Delta}$ a sequence Δ_j of pairwise disjoint open discs whose radii sum and whose union is dense in Δ. Let K be a Swiss cheese and consider $R(K)$, the uniform closure of the rational functions with poles off K. Two points $z,w \in K$ are in the same Gleason part if $\sup\{|f(w)| : f \in R(K),\ \|f\| \leq 1,\ f(z) = 0\} < 1$. The set K decomposes into $K = \bigcup_{n\geq 0} P_n$ where P_0 is the set of one point parts (peak points) and each P_n, $n \geq 1$, a non-trivial part. If ν is a measure on K which annihilates $R(K)$ and $U_{|\nu|}(z) < \infty$, $\hat{\nu}(z) \neq 0$, then z lies in some P_n, $n \geq 1$. Now let ν be $dz/2\pi i$ on $\partial\Delta$ and $-dz/2\pi i$ on each $\partial\Delta_j$. Then $\hat{\nu} = 1$ a.e. on K, so that P_0 has area zero. By the abstract F. and M. Riesz theorem, $\nu = \Sigma_{n=1}^{\infty} \nu_n$, where ν_n is orthogonal to $R(K)$, $\nu_n \perp \nu_m$ and $\hat{\nu}_n$ is the characteristic function of P_n (almost everywhere). Thus an answer to Problem 4.2 should tell us something about the structure of the parts P_n. It is not known whether $\overline{P}_n$ can meet P_m for $n,m > 0$, $n \neq m$. Details concerning the above discussion can be found in [20], [28], and [86].

An unpublished result of John Wermer answers 4.2 when the set is the interior of a Jordan curve.

<u>Theorem 4.3</u>: Let U be the region bounded by a Jordan curve Γ and assume there is a measure μ on Γ such that $\hat{\mu}(z) = 1$ for $z \in U$, $\hat{\mu}(z) = 0$ for $z \notin \Gamma \cup U$. Then Γ is rectifiable.

<u>Proof</u>: We may assume $0 \in U$. Then for any integer k we have

$$\int_\Gamma z^k d\mu = \delta_{-1,k} ,$$

so that in $C(\Gamma)$

$$\text{dist}(z^{-1}, \text{span}\{z^k : k \neq -1\}) > 0.$$

Let $\varphi : \Delta \to U$ conformally. Then φ extends to a homeomorphism of $\partial\Delta$ onto Γ and in $C(\partial\Delta)$

$$\text{dist}(\varphi^{-1}, \text{span}\{\varphi^k ; k \neq -1\}) > 0.$$

Let ν be a measure on $\partial\Delta$ such that

$$\int \varphi^k(z)d\nu(z) = \delta_{k,-1} .$$

By Mergelyan's theorem the coordinate z is in the closed algebra generated by φ on $\partial\Delta$. Hence $\int z^k d\nu(z) = 0$ for $k \geq 0$ and by the F. and M. Riesz theorem $d\nu(z) = h(z)dz$ with $h \in H^1$. We can assume that $\varphi(0) = 0$. Then for any k, and $0 < r < 1$,

$$\int_{|z|=r} \varphi^k(z)h(z)dz = \int_{|z|=1} \varphi^k(z)h(z)dz = 0 .$$

But

$$\frac{1}{2\pi i} \int_{|z|=r} \varphi^k(z)\varphi'(z)dz = \delta_{k,-1} .$$

Hence $(h(z) - \varphi'(z)/2\pi i)dz$ annihilates all integer powers of φ, so that $h(z) = \varphi'(z)/2\pi i$. Therefore $\varphi' \in H^1$, which means that Γ is rectifiable.

CHAPTER III. HAUSDORFF MEASURE

§1. Definition and Fundamentals

In this section we prove the general results about Hausdorff measures that we will need later. More information can be found in [14] and [69].

A measure function is an increasing continuous function $h(t)$, $t \geq 0$, such that $h(0) = 0$. Let E be a bounded set and for $\delta > 0$ write

$$\Lambda_h^{(\delta)}(E) = \inf\left\{ \sum_{j=1}^{\infty} h(\delta_j) : E \subset \bigcup \Delta(a_j, \delta_j);\ \delta_j \leq \delta \right\} .$$

Then $\Lambda_h^{(\delta)}(E)$ is a decreasing function of δ so that the limit

$$\Lambda_h(E) = \lim_{\delta \to 0} \Lambda_h^{(\delta)}(E)$$

exists; it is called the Hausdorff measure of E relative to h. For example, if $h(t) = t^2$, then $\Lambda_h(E)$ is essentially the (outer) area of E, and if $h(t) = 2t$ and E lies on a rectifiable curve, then $\Lambda_h(E)$ is the (outer) arc length of E, which we denote by $\ell(E)$. When $h(t) = t^r$, Λ_h is called r-dimensional Hausdorff measure, and denoted by Λ_r. If $\Lambda_1(E) < \infty$ we say E is of finite length.

The relation between Λ_1 and Painlevé length κ is given by the following theorem.

Theorem 1.1: If E is any set

$$\kappa(E) \leq 2\pi\Lambda_1(E) .$$

If E is a compact, nowhere dense set with a connected complement, then

$$\Lambda_1(E) \leq 4\kappa(E) .$$

Proof: The first assertion is clear. To prove the second let $\varepsilon > 0$, $\delta > 0$ and take a regular neighborhood V of E so that

$$E \subset V \subset E + \Delta(0,\delta/2)$$

and $\ell(\partial V) \leq (1 + \varepsilon)\kappa(E)$. Write $\partial V = \bigcup_{j\leq n} \Gamma_j$. Because $\mathbb{C}\setminus E$ is connected, we can assume the curve Γ_j surrounds an open-closed subset E_j of E and $E_j \cap E_k = \emptyset$, if $j \neq k$. Let $\delta_j = \max_{E_j} \operatorname{dist}(z,\Gamma_j)$. Then $\delta_j \leq \min(\ell(\Gamma_j),\delta/2)$. We can cover Γ_j by discs $\Delta(\zeta,\delta_j)$, $\zeta \in \Gamma_j$ using no more than

$$p_j = \left[\frac{\ell(\Gamma_j)}{\delta_j}\right] + 1$$

discs. Replacing each disc by one twice as large, we have a covering of E_j by p_j discs of radius $2\delta_j \leq \delta$. Thus

$$\Lambda_1^{\delta}(E) \leq 2 \sum_j p_j\delta_j$$

$$\leq 4 \sum \ell(\Gamma_j) \leq 4(1 + \varepsilon)\kappa(E) .$$

Clearly there is a similar result if each component of E divides the plane into at most m components. However, some restriction of this nature is necessary. A Swiss cheese is a compact set obtained by deleting from the closed unit disc pairwise disjoint open discs $\Delta(a_j,r_j)$

such that $\Sigma r_j < \infty$. Such a set has positive area, and thus infinite one dimensional Hausdorff measure, but it has finite Painlevé length.

Of course Λ_h depends only on the behavior of $h(t)$ near the origin. The next lemma follows directly from the definition.

<u>Lemma 1.2</u>: Let $h(t)$ and $H(t)$ be measure functions. Then for any bounded set E

$$\Lambda_h(E) \leq \left(\overline{\lim_{t\to 0}} \frac{h(t)}{H(t)} \right) \Lambda_H(E) .$$

Lemma 1.2 clearly implies that if $0 < r < s$ and $\Lambda_r(E) < \infty$, then $\Lambda_s(E) = 0$. The unique number $\alpha \geq 0$ such that $\Lambda_s(E) = 0$ for $s > \alpha$ and $\Lambda_r(E) = \infty$ for $r < \alpha$ is called the <u>Hausdorff dimension</u> of E.

Lemma 1.2 also implies that little is lost if we consider only smooth measure functions. Let $H(t)$ be a measure function. Since C^∞ increasing functions are dense in the continuous increasing functions on any closed interval, it is not hard to find another measure function $h(t)$ such that

$$\lim_{t\to 0} \frac{H(t)}{h(t)} = 1 . \tag{1.1}$$

$$h \text{ is } C^\infty \text{ on } t > 0 . \tag{1.2}$$

By (1.1), and Lemma 1.2 $\Lambda_h = \Lambda_H$. We will therefore assume that a measure function satisfies (1.2).

The measure Λ_h is a Caratheodory outer measure [38, p. 53], but $\Lambda_h(E)$ is frequently infinite, and thus not suitable for an upper bound. We now introduce two other set functions, which, while not additive, will be more useful.

Let $S(z;\delta)$ be the closed square with center z and side δ, having its sides parallel to the axes. For $h(t)$ a measure function and E a bounded set, we define the Hausdorff content as

$$M_h(E) = \inf\left\{\sum_{j=1}^{\infty} h(\delta_j) : E \subset \bigcup S(a_j,\delta_j)\right\}.$$

The next lemma follows easily from the definitions.

Lemma 1.3: $M_h(E) = 0$ if and only if $\Lambda_h(E) = 0$.

Now let $\mathcal{G}_n$ be the grid of squares $S(z,2^{-n})$ with center $z = (p+qi)2^{-n-1}$, p, q, odd integers, so that $\mathcal{G}_n$ is obtained from all lines $x = a2^{-n}$, $y = b2^{-n}$, a, b, integers. Then $\mathcal{G} = \bigcup_{n=-\infty}^{\infty} \mathcal{G}_n$ is the grid introduced at the end of §2 of II. Define

$$m_h(E) = \inf\left\{\sum_{j=1}^{\infty} h(\delta_j) : E \subset \bigcup S(a_j,\delta_j);\ S(a_j,\delta_j) \in \mathcal{G}\right\}.$$

Since $S(a,\delta)$ can be covered by nine squares from some $\mathcal{G}_n$ with $2^{-n} \le \delta$, we have

Lemma 1.4: For any measure function h and any bounded set E,

$$M_h(E) \le m_h(E) \le 9M_h(E) .$$

Consequently Λ_h, M_h and m_h have the same null sets.

It is obvious that

$$M_h(E) = \inf\{M_h(U) : U^{\text{open}} \supset E\} .$$

If $\lim_{t\to 0} h(t)/t = 0$, so that lines have measure zero, the same holds for m_h:

$$m_h(E) = \inf\{m_h(U) : U^{\text{open}} \supset E\} .$$

Under the same hypothesis on $h(t)$, we have

$$m_h(E_n) \nearrow m_h(E) \quad \text{if} \quad E_n \nearrow E .$$

This means m_h is a capacity in the sense of Choquet:

$$m_h(E) = \sup\{m_h(F) : F^{\text{compact}} \subset E\}$$

for any analytic set E. See [14] for the proofs of these facts.

We will need the following theorem of Frostman [27]. See also [14] and [54]. To simplify the proof, we assume the measure function satisfies

$$\lim_{t\to 0} \frac{h(t)}{t} = 0 ,$$

although the theorem is true, with a smaller constant, without this hypothesis. We say a measure μ is of <u>growth</u> $h(t)$ if

$$|\mu|(\Delta(z,\delta)) \leq h(\delta)$$

for all z and δ.

<u>Theorem 1.5</u> (Frostman): Let $h(t)$ be a measure function such that $\lim_{t\to 0} h(t)/t = 0$, and let E be a bounded set. If σ is a positive

measure of growth $h(t)$, then

$$\sigma(E) \leq 4M_h(E) .$$

Conversely, if E is an analytic set, there is a positive measure σ of growth $h(t)$ such that

$$\sigma(E) \geq \frac{m_h(E)}{25} .$$

Proof: Since $\sigma(S(z,\delta)) \leq 4h(\delta)$ if σ has growth $h(t)$, the first assertion is trivial. In proving the second we assume that E is compact. For $n \geq 0$ define a measure ν_n first such that for each $S \in \mathcal{G}_n$, $\nu_n(S) = h(2^{-n})$ if $S \cap E \neq \emptyset$ and $\nu_n(S) = 0$ if $S \cap E = \emptyset$. If for some $S \in \mathcal{G}_{n-1}$, $\nu_n(S) > h(2^{-n+1})$ reduce the masses on the squares from $\mathcal{G}_n$ inside S until $\nu_n(S) = h(2^{-n-1})$. Repeat this for all squares in $\mathcal{G}_{n-2}, \mathcal{G}_{n-3}, \ldots$, and let μ_n be the resulting measure. Then $\mu_n(S) \leq h(\delta)$ if $S \in \mathcal{G}$ has side $\delta \geq 2^{-n}$, and $m_h(E) \leq \|\mu_n\| \leq h(\delta_0)$, where $E \subset S(z,\delta_0) \in \mathcal{G}$. Let μ be a weak-star limit of $\{\mu_n\}$. Then μ has support E and $\mu(E) \geq m_h(E)$. Since $\mu(V) \leq \overline{\lim}\, \mu_n(V)$ for every open set V and since $\lim_{t\to 0} h(t)/t = 0$, $\mu(\partial S) = 0$ for all $S \in \mathcal{G}$. Therefore $\mu(S) \leq h(\delta)$ if $S \in \mathcal{G}$ has side δ. Hence $\mu(\Delta(z,\eta)) \leq 25h(\eta)$ and $\sigma = \mu/25$ has growth $h(t)$.

An important fact which we will not use, and therefore have not proved, is that every analytic set of positive Hausdorff measure contains a compact set of positive finite Hausdorff measure. See, for example, [14] for a proof.

§2. Removing Singularities

Hausdorff measure can be used to give sufficient conditions for a set E to have analytic capacity zero or for a function on $S^2 \setminus E$ to be a Cauchy transform. The simplest result of this type is Painlevé's theorem, proved in the introduction. In fact, all the proofs in this section are little more than variations of that argument.

<u>Theorem 2.1</u>: If $\Lambda_1(E) < \infty$ and $f \in A(E,m)$, then there is a measure μ on E such that

$$\|\mu\| \leq m\Lambda_1(E)$$

and

$$f(z) = \hat{\mu}(z) \quad \text{a.e.}$$

Moreover,

$$\gamma(E) \leq \Lambda_1(E) .$$

<u>Proof</u>: Let F be a compact subset of E such that $f \in A(F,m)$. Then F has area zero and Painlevé length at most $2\pi\Lambda_1(E)$. The theorem now follows Theorem 3.1 of Chapter II.

The <u>modulus of continuity</u> of a continuous function f is denoted by

$$\omega_f(\delta) = \sup\{|f(z) - f(w)| : |z - w| < \delta\} .$$

If $f \in C(E,m)$ for some bounded set E, then $\omega_f(\delta)$ is well defined because the corresponding supremum is attained for z and w near E. Because

$$(2.1) \qquad \left| \int_{\partial S(a,\delta)} f(z)dz \right| = \left| \int_{\partial S(a,\delta)} (f(z) - f(a))dz \right|$$

$$\leq 4\delta\omega_f(\delta) ,$$

the measure function $t\omega_f(t)$ is a natural one to use when discussing analytic functions. Indeed we have:

Lemma 2.2: Let E be a closed subset of an open set Ω and let f be continuous on $\overline{\Omega}$ and analytic on $\Omega \setminus E$. Then for any square S in the grid $\mathcal{G}$ with $S \subset \overline{\Omega}$, we have

$$(2.2) \qquad \left| \int_{\partial S} f(z)dz \right| \leq 4m_h(S \cap E) ,$$

where $h(t) = t\omega_f(t)$.

Proof: For $\varepsilon > 0$ there is a cover $(S_j)_{j=1}^{\infty}$ of $S \cap E$ with $S_j \subset S$, $S_j \in \mathcal{G}$, $S_j^0 \cap S_k^0 = \emptyset$ and

$$\sum_j h(\delta_j) \leq m_h(S \cap E) + \varepsilon .$$

Then

$$\int_{\partial S} f(z)dz = \sum_j \int_{\partial S_j} f(z)dz$$

so that by (2.1)

$$\left|\int_{\partial S} f(z)dz\right| \leq 4 \sum_j h(\delta_j) \leq 4m_h(S \cap E) + 4\varepsilon \ .$$

Theorem 2.3: Let E be a closed subset of an open set Ω and let $f \in C(\overline{\Omega})$ be analytic on $\Omega\setminus E$. Let $h(t) = t\omega_f(t)$. If $m_h(E) = 0$, then f is analytic on Ω.

Proof: By Lemma 2.2 and the continuity of f we have $\int_{\partial S} f(z)dz = 0$ for any square $S \subset \overline{\Omega}$. Morera's theorem now tells us f is analytic on Ω.

Corollary 2.4: If E is a countable union of sets of finite length, then $\alpha(E) = 0$.

Proof: Assume $f \in C(E,m)$. Let $h(t) = t\omega_f(t)$. By Lemma 1.2, $m_h(E) = 0$ so that f extends analytically across E, and $f = 0$ by Liouville's theorem.

Theorem 2.5: Let $f \in C(E,m)$ and let $h(t) = t\omega_f(t)$. If $\Lambda_h(E) < \infty$, there is a measure μ on E such that $\hat{\mu}(z) = f(z)$ a.e. and $\|\mu\| \leq \frac{18}{\pi} \Lambda_h(E)$.

Proof: Let K be a compact subset of E such that $f \in C(K,m)$. Then $\Lambda_h'(K) \leq \Lambda_h(E)$. Let

$$m_h^{(n)}(K) = \inf\left\{\sum h(\delta_j) : F \subset \bigcup S(a_j,\delta_j),\ S(a_j,\delta_j) \in \mathcal{G},\ \delta_j \leq 2^{-n}\right\}.$$

We have

$$\frac{1}{9}\, m_h^{(n)}(K) \leq \Lambda_h^{(2^{-n})}(K) \leq m_n^{(n)}(K)$$

because every disc of radius less than 2^{-n} can be covered by 9 squares from $\mathcal{G}$ of side of most 2^{-n}, while every such square lies in a disc of radius less than 2^{-n}. Let $S_1, S_2, \ldots, S_p$ be a covering of K by squares from $\mathcal{G}$ with sides δ_j at most 2^{-n}. Then

$$\sum \left| \frac{1}{2\pi i} \int_{\partial S_j} f(z)dz \right| \leq \sum_j \frac{2}{\pi} \delta_j \omega_f(\delta_j)$$

$$\leq \frac{2}{\pi} m_h^{(n)}(K) \ .$$

Hence

$$V_{\mathcal{G}}(f) \leq \frac{2}{\pi} \lim_{n\to\infty} m_h^{(n)}(K) \leq \frac{18}{\pi} \Lambda_h(E) \ ,$$

and by Corollary 2.6 of II there is a measure μ on K such that $\hat{\mu} = f$ a.e. and $\|\mu\| = V_{\mathcal{G}}(f) \leq \frac{18}{\pi} \Lambda_h(E)$.

§3. Newtonian Capacity

Since the absolute value of the Cauchy kernel is Newton's kernel $|z - w|^{-1}$, it is convenient to discuss absolutely convergent Cauchy transforms in terms of Newtonian potential theory. While there are many treatments of potential theory in print, [10], [14], [49], [77], we only need two basic theorems and so include their proofs.

Let $K(z) = 1/|z|$ be Newton's kernel. The Newtonian potential of any positive compactly supported measure σ is the convolution

$$U_\sigma(z) = K * \sigma(z) = \int \frac{1}{|z - w|} d\sigma(w) \ .$$

For any bounded set E we define the set of measures

$$\Gamma(E) = \{\sigma : \sigma > 0,\ S_\sigma \subset E \text{ and } U_\sigma \leq 1 \text{ almost everywhere } \sigma\} .$$

A measure σ lies in $\Gamma(E)$ if and only if $S_\sigma \subset E$ and $U_\sigma \leq 1$ everywhere. This is a consequence of the following

<u>Theorem 3.1</u>: (Maximum Principle). If $U_\sigma \leq M$ almost everywhere σ, then $U_\sigma \leq M$ everywhere.

<u>Proof</u>: Let $\varepsilon > 0$, and assume $U_\sigma \leq M$ a.e. σ. By Egoroff's theorem there is a compact subset F of S_σ such that

(i) $U_\sigma \leq M$ on F

(ii) $\sigma(\mathbb{C}\setminus F) < \varepsilon$

(iii) $\lim_{\delta\to 0} \sup_{z\in F} \int_{\Delta(z,\delta)} \frac{d\sigma(w)}{|w - z|} = 0.$

Conditions (i)-(iii) hold for the restricted measure $\sigma|F$ and it is enough to prove the assertion for $\sigma|F$. Indeed, let $\varepsilon_n \searrow 0$ and let $\{F_n\}$ be an increasing sequence of compact sets such that (i)-(iii) hold for F_n and ε_n. If $\sigma_n = \sigma|F_n$ and $U_{\sigma_n} \leq M$, then by monotone convergence $U_\sigma \leq M$.

So we will replace σ by $\sigma|F$. Now choose δ so that

$$\int_{\Delta(z,2\delta)} \frac{d\sigma(w)}{|w - z|} < \varepsilon$$

for all $z \in F$. Because U_σ is subharmonic off F we will be done if we show

$$\overline{\lim_{F\ni z\to a}}\ U_\sigma(z) \leq M$$

for every $a \in F$. Now

$$\overline{\lim_{F \not\ni z \to a}} U_\sigma(z) = \int_{F \setminus \Delta(a,\delta)} \frac{d\sigma(w)}{|w - a|} + \overline{\lim_{F \not\ni z \to a}} \int_{\Delta(a,\delta)} \frac{d\sigma(w)}{|w - z|} .$$

The first integral is bounded by M. To estimate the other, fix $z \in \Delta(a,\delta) \setminus F$ and take six closed cones $Q_1, \ldots, Q_6$ with vertex z and angle $\pi/6$. Let $w_\nu \in Q_\nu \cap F$ be closest to z. Then $|w - w_\nu| \leq |w - z|$ for $w \in Q_\nu \cap F$. Therefore

$$\int_{Q_\nu \cap \Delta(a,\delta)} \frac{d\sigma(w)}{|w - z|} \leq \int_{\Delta(w_\nu, 2\delta)} \frac{d\sigma(w)}{|w - w_\nu|} < \varepsilon$$

so that

$$\int_{\Delta(a,\delta)} \frac{d\sigma(w)}{|w - z|} < 6 \cdot \varepsilon .$$

This bounds the last limes superior by 6ε, and so we have $U_\sigma \leq M + 6\varepsilon$ everywhere, which proves the theorem.

Write $K_n(z) = \min\{K(z), n\}$, and set

$$U_\sigma^{(n)}(z) = K_n * \sigma(z) = \int K_n(z - w) d\sigma(w) .$$

The $U_\sigma^{(n)}$ are continuous and increase to U_σ, which is consequently lower semi-continuous.

<u>Theorem 3.2</u>: If σ is a positive measure and F is a compact set, then the following are equivalent:

(i) U_σ is continuous on F

(ii) $U_\sigma^{(n)} \to U_\sigma$ uniformly on F

(iii) $\lim_{\delta\to 0} \sup_{z\in F} \int_{\Delta(z,\delta)} \frac{d\sigma(w)}{|w - z|} = 0.$

Proof: It is clear that (iii) $\Rightarrow$ (ii) $\Rightarrow$ (i). If (i) holds, then for any z and δ, $\sigma(\partial\Delta(z,\delta)) = 0$ so that

$$\int_{C\setminus\Delta(z,\delta)} \frac{d\sigma(w)}{|w - z|}$$

is a continuous function at z. But by (i) this means

$$\int_{\Delta(z,\delta)} \frac{d\sigma(w)}{|w - z|}$$

is continuous. By (i) no point of F has positive measure, so that (iii) follows via Dini's theorem.

Using (iii) we see that the proof of 3.1 actually yielded the following.

Theorem 3.3: U_σ is continuous on S^2 if U_σ is continuous on the support S_σ of σ.

Theorem 3.4: Let μ be a finite complex measure with compact support and assume $U_{|\mu|}$ is continuous. Then the Cauchy transform $\hat{\mu}$ is continuous.

Proof: Let C_n be continuous with $|C_n| \le n$ and $C_n(z) = 1/z$ if $|z| \ge 1/n$. Then

$$C_n * \mu(z) = \int C_n(z - w)d\mu(w)$$

is continuous and

$$|C_n * \mu(z) - \hat{\mu}(z)| \leq U_{|\mu|}(z) - U^{(n)}_{|\mu|}(z) ,$$

which converges uniformly to zero by 3.2 and 3.3.

The <u>Newtonian capacity</u> of E is

$$C(E) = \sup\{\sigma(E) : \sigma \in \Gamma(E)\} .$$

By Egoroff's theorem and 3.2 we have

$$C(E) = \sup\{\sigma(E) : \sigma \in \Gamma(E) \text{ and } U_\sigma \text{ is continuous}\} .$$

In light of this we have

<u>Corollary 3.5</u>: For any bounded set E,

$$C(E) \leq \alpha(E) \leq \gamma(E) .$$

<u>Corollary 3.6</u>: If E is a countable union of sets of finite length, then $C(E) = 0$.

This is a variant of the Erdös-Gillis theorem [14], [77].

The following exercise shows that the converse of 3.4 fails.

<u>Exercise 3.7</u>: (a) Partition the unit square into two sets A_n and B_n where

$$A_n = \{k/n < x \leq (k + 1)/n : k \text{ even}\}$$

$$B_n = \{k/n < x \leq (k + 1)/n : k \text{ odd}\} .$$

Let μ_n be absolutely continuous with density 1 on A_n and -1 on

B_n. Prove $U_{|\mu_n|}$ is continuous but that $\|\hat{\mu}_n\| \to 0$ $(n \to \infty)$.

(b) Let $a_k \searrow 0$ and let $S_k = S(a_k, \delta_k)$ where δ_k is very small. Choose measures ν_k on S_k such that $\|\nu_k\| = \delta_k$ and ν_k looks like μ_n for sufficiently large $n = n(k)$. Let $\nu = \Sigma\, \nu_k$. Prove that for suitable δ_k and $n(k)$, $U_{|\nu|} \leq 1$, $U_{|\nu|}$ is not continuous, but $\hat{\nu}$ is continuous.

Constructions like this occur quite often; they are all instances of the vague principle that if a property of sets fails to hold uniformly then there is a set for which it fails altogether.

Exercise 3.8: Let E be a compact set with $C(E) > 0$. Prove there is $f \in C(E,1)$ which is not almost everywhere equal to a Cauchy transform. Hint: Let E be totally disconnected. Produce a sequence $f_n = \hat{\mu}_n \in C(E,1)$ such that $\mu_n \in \Gamma(E)$, $\{f_n\}$ is uniformly convergent, but $\|\mu_n\| \to \infty$. This can be done by taking $\mu \in \Gamma(E)$ and successive partitions $E_1^{(n)}, \ldots, E_{m_n}^{(n)}$ of E such that if $\nu_n = (-1)^k \mu$ on $E_k^{(n)}$ then $\|\hat{\nu}_n\|_\infty < 2^{-n}$. Then $f_n = \Sigma_{j \leq n}\, \hat{\nu}_j$ will have the desired properties. See [80] for more details.

§4. Hausdorff Measure and Cauchy Transforms

In this section we give some metric conditions which ensure that $C(F) > 0$ and thus that $\alpha(F) > 0$. The idea here goes back to Frostman [27]. With a little more effort we can even exhibit $f \in C(F,m)$ with an estimate on $\omega_f(\delta)$.

Let $h(t)$ be a measure function satisfying the integrability condition

$$(*) \qquad \int_0 \frac{h(t)}{t^2}\, dt < \infty .$$

Examples of such measure functions include $h(t) = t^\beta$, $\beta > 1$ and $h(t) = t/(\log 1/t)^\beta$, $\beta > 1$. Let σ be a positive measure of growth $h(t)$, and write $m(z,r) = \sigma(\Delta(z;\, r))$. Then $m(z,r) \leq \min(h(r), \|\sigma\|)$, and so $m(z,r)/r$ has limit 0 at ∞. Also, as

$$\frac{m(z,r)}{r} \leq \frac{h(r)}{r} \leq \frac{1}{2} \int_r^{2r} \frac{h(t)}{t^2}\, dt ,$$

$m(z,r)/r$ has limit 0 at 0. Integrating by parts now gives

$$U_\sigma(z) = \int \frac{d\sigma(w)}{|w - z|} = \int_0^\infty \frac{dm(z,r)}{r} = \int_0^\infty \frac{m(z,r)}{r^2}\, dr ,$$

where the middle integral is a Riemann-Stieltjes integral. Defining R by $h(R) = \|\sigma\|$, we have

$$(4.1) \qquad U_\sigma(z) \leq \int_0^R \frac{h(r)}{r^2}\, dr + \frac{\|\sigma\|}{R} = \int_0^R \frac{dh(r)}{r} .$$

Combined with Frostman's theorem (Theorem 1.5) this estimate shows $C(E) > 0$ whenever $m_h(E) > 0$ and h satisfies $(*)$. More precisely, we have:

Theorem 4.1: Let $h(t)$ be a measure function satisfying $(*)$. Then for any analytic set E

$$(4.2) \qquad C(E) \geq \frac{m_h(E)}{25 \int_0^R \frac{dh(r)}{r}}$$

where $h(R) = m_h(E)/25$.

Proof: By 1.5 there is a positive measure σ on E having growth $h(t)$ such that $\|\sigma\| \geq m_h(E)/25$. Using the estimate (4.1) we find that

$$\frac{1}{C(E)} \leq \frac{\|U_\sigma\|_\infty}{\|\sigma\|} \leq \frac{1}{h(R)} \int_0^R \frac{dh(r)}{r} ,$$

as required.

On p. 34 of [14] it is shown that there exist sets E with $C(E) > 0$ but $M_h(E) = 0$ for all h satisfying (*). See also [13].

When $h(t) = t^{1+\beta}$, $0 < \beta \leq 1$ we have

$$\int_0^R \frac{dt^{1+\beta}}{t} = \frac{\beta + 1}{\beta} R^\beta$$

and (4.2) has a particularly simple form.

Corollary 4.2: If $0 < \beta \leq 1$ and if E is an analytic set, then

$$\alpha(E) \geq C(E) \geq \frac{\beta}{1 + \beta} \left(\frac{m_{1+\beta}(E)}{25} \right)^{1/1+\beta} .$$

For $\beta = 1$ this was proved in the introduction by essentially the same argument.

Corollary 4.3: Let ν be a positive compactly supported measure. If $h(t)$ satisfies (*) then

$$m_h(\{z : U_\nu(z) = \infty\}) = 0$$

and

$$C(\{z : U_\nu(z) = \infty\}) = 0 \ .$$

In particular, U_ν converges except on a set of Hausdorff dimension at most one.

<u>Proof</u>: Because U_ν is lower semi-continuous,

$$E = \{z : U_\nu(z) = \infty\}$$

is a G_δ-set. Assume $m_h(E) > 0$. Then there is a compact $F \subset E$ and a measure σ on F of growth $h(t)$, such that U_σ is bounded. But then

$$\infty = \int U_\nu d\sigma = \int U_\sigma d\nu < \infty \ ,$$

a contradiction. Of course the above argument really shows that $U_\nu < \infty$ except on a set of capacity zero.

Theorem 4.1 also has as a corollary

$$\alpha(E) \geq \frac{m_h(E)}{25 \int_0^R \frac{dh(r)}{r}}$$

for $h(t)$, E and R as above. However, following Dolzhenko [23], we can get more information.

<u>Theorem 4.4</u>: Assume $h(t)$ satisfies (*) and let F be a compact set with $m_h(F) > 0$. Let

$$h(R) = \frac{m_h(F)}{25} \ ,$$

and let

$$M = \left(\int_0^R \frac{dh(t)}{t} \right)^{-1} .$$

Then there exists $f \in C(F,M)$ such that $f'(\infty) = h(R)$ and

$$(4.3) \qquad \omega_f(\delta) \le 4 \int_0^{\delta/2} \frac{dh(r)}{r} + 4\delta \int_{\delta/2}^R \frac{dh(r)}{r^2} .$$

The right hand side of (4.3) tends to zero with δ. For, fixing η such that $\int_0^\eta \frac{dh(r)}{r} < \varepsilon$, we have for $\delta < \eta$

$$4\delta \int_{\delta/2}^R \frac{dh(r)}{r^2} \le 8 \int_{\delta/2}^\eta \frac{dh(r)}{r} + 4\delta \int_\eta^R \frac{dh(r)}{r^2} .$$

Proof: Take a measure σ on F of growth $h(t)$ such that $\sigma(F) = h(R)$, and let $f(z) = \hat{\sigma}(z)$. Then $\|f\| \le M$ and $f'(\infty) = h(R)$. To verify the estimate (4.3), which also shows that $f \in C(E,M)$, we take z_1 and z_2 and write $\delta = |z_1 - z_2|$. Then

$$|f(z_1) - f(z_2)| \le \delta \int \frac{d\sigma(w)}{|w - z_1||w - z_2|} .$$

We integrate over four sets

$$A = \{|w - z_1| < \delta/2\}$$
$$B = \{|w - z_2| < \delta/2\}$$
$$C = \{|w - z_1| \le |w - z_2|,\ |w - z_1| \ge \delta/2\}$$
$$D = \{|w - z_2| < |w - z_1|,\ |w - z_2| \ge \delta/2\} .$$

Now

$$\delta \int_A \frac{d\sigma(w)}{|w - z_1||w - z_2|} \leq 2 \int_A \frac{d\sigma(w)}{|w - z_1|}$$

$$= 2 \int_0^{\delta/2} \frac{dm(z_1,r)}{r} = \frac{4m(z_1,\delta/2)}{\delta} + 2 \int_0^{\delta/2} \frac{m(z_1,r)dr}{r^2}$$

$$\leq \frac{4m(z_1,\delta/2)}{\delta} + 2 \int_0^{\delta/2} \frac{h(r)}{r^2}\, dr \ .$$

Similarly

$$\delta \int_B \frac{d\sigma(w)}{|w - z_1||w - z_2|} \leq \frac{4m(z_2,\delta/2)}{\delta} + 2 \int_0^{\delta/2} \frac{h(r)}{r^2}\, dr \ .$$

Also

$$\delta \int_C \frac{d\sigma(w)}{|w - z_1||w - z_2|} \leq \delta \int_C \frac{d\sigma(w)}{|w - z_1|^2} \leq \delta \int_{\delta/2}^{\infty} \frac{dm(z_1,r)}{r^2}$$

$$= 2\delta \int_{\delta/2}^{\infty} \frac{m(z_1,r)}{r^3}\, dr - \frac{4m(z_1,\delta/2)}{\delta}$$

$$\leq 2\delta \int_{\delta/2}^{R} \frac{h(r)}{r^3}\, dr + \frac{\delta h(R)}{R^2} - \frac{4m(z_1,\delta/2)}{\delta} \ .$$

There is a similar estimate for the integral over D. Adding, we have

$$|f(z_1) - f(z_2)| \leq 4 \int_0^{\delta/2} \frac{h(r)}{r^2}\, dr + 4\delta \int_{\delta/2}^{R} \frac{h(r)}{r^3}\, dr + \frac{2\delta h(R)}{R^2}$$

$$= 4 \int_0^{\delta/2} \frac{dh(r)}{r} + 4\delta \int_{\delta/2}^{R} \frac{dh(r)}{r^2} \ .$$

Again, when $h(t) = t^{1+\beta}$, $0 < \beta < 1$, the expressions simplify, because

$$4\int_0^{\delta/2} \frac{dh(r)}{r} + 4\delta \int_{\delta/2}^R \frac{dh(r)}{r^2} = 4(1+\beta)\int_0^{\delta/2} r^{\beta-1}dr + 4\delta(1+\beta)\int_{\delta/2}^R r^{\beta-2}dr$$

$$\leq 4(1+\beta)\left(\frac{2^{-\beta}}{\beta} + \frac{2^{1-\beta}}{1-\beta}\right)\delta^\beta ,$$

and the corresponding Cauchy transform is in the Hölder class Lip β.

Corollary 4.5: Let $0 < \beta < 1$. If E is an analytic set and $m_{1+\beta}(E) > 0$ then there is a constant C_β and a function $f \in C(E,1)$ such that $f'(\infty) \geq C_\beta(m_{1+\beta}(E))^{1/1+\beta}$ and $|f(z) - f(w)| \leq C_\beta^{-1} |z - w|^\beta$. Conversely, if E is a relatively closed subset of a bounded open set Ω, and $m_{1+\beta}(E) = 0$, then every $f \in \mathrm{Lip}_\beta(\overline{\Omega}\setminus E)$ which is analytic on $\Omega\setminus E$ extends to be analytic on Ω. The constant C_β is asymptotic to β as β tends to 0.

Proof: The first assertion holds by 4.2 and the remark following 4.4. The converse is a consequence of 2.3 once we observe that E is nowhere dense and that each $f \in \mathrm{Lip}_\beta(\overline{\Omega}\setminus E)$ has an extension in $\mathrm{Lip}_\beta(\overline{\Omega})$.

It seldom happens that a metric condition as in 4.5 is both necessary and sufficient for a singularity set to be removable for a class of analytic functions. The reader may have noticed the similarity between 4.5 and Privalov's theorem: if u is in Lip_β on $[-\pi,\pi]$ and $\beta < 1$, then its conjugate function is in Lip_β. See [67], [91]. In fact (4.3) is essentially the same as the estimate for the modulus of continuity of $^*u$ given on p. 121 of [91]. Of course this is no coincidence; in both cases the kernel has singularity in the order of $1/r$.

When the Hausdorff measure is area, the estimate for $\alpha(E)$ given in 4.1 can be refined somewhat. We can take σ to be the area measure, and thus remove the factor of $1/25$. However an even better estimate is available, which is sharp in the case of a disc.

Theorem 4.6: (Ahlfors-Beurling [5]). For any analytic set E,

$$\alpha(E) \geq \left(\frac{\text{Area}(E)}{\pi} \right)^{1/2} .$$

Proof: Let $f(z) = \iint_F \frac{dudv}{u + iv - z}$ where F is a compact subset of E. Translate F so that $\|f\| = |f(0)|$ and then rotate F so that $f(0) = \|f\|$. We now estimate

$$f(0) = \iint_F \frac{dudv}{u + iv} = \iint_F \cos\theta \, drd\theta .$$

Let $F^+ = F \cap \{u > 0\}$, and let $\ell(r,\theta)$ be the length of the set of points $se^{i\theta} \in F^+$ with $s \leq r$. Then

$$\iint_F \frac{dudv}{u - iv - z} \leq \iint_{F^+} \cos\theta \, drd\theta$$

$$= \int_{-\pi/2}^{\pi/2} \ell(\infty,\theta)\cos\theta \, d\theta \leq \left(\frac{\pi}{2} \int_{-\pi/2}^{\pi/2} (\ell(\infty,\theta))^2 d\theta \right)^{1/2} .$$

But $r - \ell(r,\theta)$ is increasing and positive, so that

$$\int r \, dr \geq \int \ell(r,\theta) d\ell(r,\theta) = \frac{(\ell(\infty,\theta))^2}{2} .$$

Thus

$$\text{Area}(F^+) = \iint_{F^+} r\,dr d\theta \geq \int_{-\pi/2}^{\pi/2} \frac{(\ell(\infty,\theta))^2}{2}\,d\theta$$

and so

$$|f(z)| \leq \iint_F \frac{dudv}{u + iv - z} \leq \pi(\text{Area } F)^{1/2} .$$

Since $|f'(\infty)| = \text{Area}(F)$ the required inequality for $\alpha(F)$ holds.

Exercise 4.7: Prove that the function $f(z)$ in the above proof satisfies

$$\omega_f(\delta) \leq 4\pi\delta(1 + \log(2/\delta)) .$$

Show that if $g \in L^\infty$, then

$$f(z) = \iint \frac{g(\zeta)d\xi d\eta}{\zeta - z}$$

is in the Zygmund class:

$$|f(z + h) + f(z - h) - 2f(z)| = O(h) .$$

This, and the continuity of f, implies the estimate on $\omega_f(\delta)$ above.

Problem 4.8: Characterize the set of measures whose Cauchy transforms lie in the Zygmund class. This remote possibility is suggested by the analogy with conjugate function theory, and by the theorem that the Zygmund class is self conjugate [90], [91].

Exercise 4.9: Let $0 < \beta < 1$. Let $f \in C(E,1) \cap \text{Lip}_\beta : |f(z) - f(w)| \leq c|z - w|^\beta$. If $\Lambda_{1+\beta}(E) < \infty$, show that f is a Cauchy transform. If F is an analytic set such that $\Lambda_{1+\beta}(F) = 0$, show that f is in the uniform closure of $C(E\setminus F,1)$. Prove the same approximation theorem assuming $\Lambda_{1+\beta}(F) < \infty$ and $|f(z) - f(w)| = o(|z - w|^{\beta})$.

§5. A Problem with Lip_1

We know from §2 (or even Exercise 2.7 of II) that a set E of zero area is removable for analytic functions in Lip_1. However if $\mathrm{Area}(E) > 0$, the theorems in §4 only yield a function in $C(E,1)$ which is "Lip log"

$$\omega_f(\delta) = O(\delta \log 1/\delta) .$$

Problem 5.1: Give a condition which implies $C(E,1) \cap \mathrm{Lip}_1$ is non-trivial. Trivially $\mathrm{int}(E) \neq \emptyset$ is one condition, but almost surely it can be improved upon.

If there is a constant K such that any two points in $\mathbb{C}\setminus E$ can be joined by an arc γ in $\mathbb{C}\setminus E$ with length $\ell(\gamma) \leq K|z - w|$, then $C(E,1) \cap \mathrm{Lip}_1$ coincides with the class of functions in $C(E,1)$ having a bounded derivative. These two classes are not always the same however, as the example in IV §1 indicates.

In [12] Carleson showed that a subset E of a disc Δ has a non-constant function in $C(E,1)$ with a bounded derivative provided $C_p(\Delta\setminus E) < C_p(\Delta)$, where $1 < p < 2$ and C_p is the capacity corresponding to the kernel $|z|^{-p}$. This capacitary condition is not a necessary one because the example in §1, IV above can be modified so that $C(E) = 0$, which implies that $C_p(E) = 0$ for $1 < p < 2$ and hence that $C_p(\Delta\setminus E) = C_p(\Delta)$. It seems likely that if $C_p(\Delta\setminus E) < C_p(\Delta)$ then that for any $K > 0$ there are z and w in $\mathbb{C}\setminus E$ which cannot be connected by an arc of length bounded by $K|z - w|$.

There is a third class of functions which might be relevant to the problem. This is the set of $f \in A(E,1)$ such that f has all periods zero:

$$\int_\gamma f(z)dz = 0$$

whenever γ is a closed curve in Ω. If $f \in C(E,M)$ and $|f'| < 1$, then f' is in this class. If there is a fixed z_0 in $\mathbb{C}\setminus E$ such that any $z \in \mathbb{C}\setminus E$ can be connected to z_0 by an arc γ with $\ell(\gamma) \leq K|z - z_0|$, then it is easily seen that a bounded function with periods zero has a bounded integral. Recall that Corollary 2.4 of II gives a geometric condition which implies every function in $A(E,1)$ with periods zero is constant.

CHAPTER IV. SOME EXAMPLES

§1. Carleson's Example

In this section it is shown that $\Lambda_1(E) = \infty$ is the only metric condition necessary for $\alpha(E) > 0$. It is also shown that $C(E) > 0$ is not a necessary condition. The construction comes from [12].

Theorem 1.1: Let $E = K \times L$ where K and L are compact subsets of the x and y coordinate axes respectively, such that K has positive length and L is uncountable. Let f be bounded and analytic off K and let ν be a continuous measure on L. Then

$$F(z) = \int_L f(z - is)d\nu(s)$$

converges absolutely for all z, and $F(z)$ is in $C(E, \|f\|\|\nu\|)$. Moreover $F'(\infty) = f'(\infty) \int d\nu$, and

$$F'(z) = \int f'(z - is)d\nu(s), \quad z \notin E .$$

If $f(z) = \hat{\mu}(z)$, we have $F(z) = \widehat{\mu \times \nu}(z)$.

Proof: The analyticity of F, the bound $\|F\| \leq \|f\|\|\nu\|$, and the equality $F'(\infty) = f'(\infty) \int d\nu$ are all quite routine. Fix $z_0 = x_0 + iy_0$ and let J be an open interval containing y_0 such that $|\nu|(J) < \varepsilon$. Then

$$F(z) = \int_{L \setminus J} f(z - is)d\nu(s) + \int_J f(z - is)d\nu(s) .$$

The first integral is continuous at z, while the second satisfies

$$\left|\int_J f(z - is)d\nu(s)\right| \leq \int_J \|f\| d|\nu| \leq \varepsilon\|f\| .$$

This means the integral is absolutely convergent at z_0 and that F is continuous at z_0. Now $F'(z) = \int f'(z - is)d\nu(s)$ follows by differentiating inside the integral. Regarding ν as a measure on the imaginary axis in the plane, we have $F = f * \nu$ where $*$ denotes convolution. Then in the sense of distributions

$$\frac{\partial F}{\partial \bar{z}} = \frac{\partial f}{\partial \bar{z}} * \nu ,$$

and if $f = \hat{\mu}$, then $\frac{\partial f}{\partial \bar{z}} = -\pi\mu$ so $\frac{\partial F}{\partial \bar{z}} = -\pi(\mu * \nu)$ and $F = \widehat{\mu * \nu} = \widehat{\mu \times \nu}$.

Theorem 1.2: Let $E = [0,1] \times L$, where L is a perfect compact subset of the y-axis. Then there is F in $C(E,1)$ such that each derivative $F^{(n)}$ of F extends continuously to S^2.

Proof: Let φ be C_0^∞ on $(0,1)$, and set

$$f(z) = \int_0^1 \frac{\varphi(t)}{t - z} dt .$$

Since φ is smooth f is bounded. An integration by parts gives

$$f^{(n)}(z) = \int_0^1 \frac{\varphi^{(n)}(t)}{t - z} dt ,$$

which is also bounded. Let ν be a continuous measure on L. Then

$$F(z) = \int f(z - is)d\nu(s)$$

is in $C(E,1)$ if $\|f\|$ is sufficiently small, and $F^{(n)}(z)$ has a continuous extension to S^2 for all n, by Theorem 1.1.

Notice that if γ is a curve in $\mathbb{C}\setminus E$, then $\int_\gamma F^{(n)}(z)dz = 0$ for $n \geq 1$. This shows that Corollary 2.4 of II is sharp. It also shows that no improvement can be made by considering "higher order periods," because

$$\int_\gamma F^{(n)}(z)z^k dz = 0, \quad 0 \leq k < n .$$

Theorem 1.3: Let $h(t)$ be a measure function such that $\lim_{t\to 0} h(t)/t = 0$. Then there is a compact set E and a measure τ on E such that

(i) $M_h(E) = 0$ and $C(E) = 0$

(ii) $\alpha(E) > 0$

(iii) $\hat{\tau}$ and all its derivatives extend continuously to E.

Proof: Let L be a perfect subset of the y-axis and let $E = [0,1] \times L$. Let φ be in $C_o^\infty(0,1)$ and let ν be a continuous measure on L. Setting $\tau = \varphi(t)dt \times \nu$ we have (ii) and (iii) by Theorems 1.1 and 1.2.

Let $H(t) = h(t)/t$. We can find a perfect set L such that $M_H(L) = 0$, because $H(t) \to 0$ $(t \to 0)$. It is then easily verified that $M_h(E) = 0$. At the same time L can be chosen to have zero logarithmic capacity. Then $C(E) = 0$ because of the following lemma.

Lemma 1.4: Let L be a compact subset of the real line. Then $C([0,1] \times L) > 0$ if and only if L has positive logarithmic capacity.

Proof: Assume $C([0,1] \times L) > 0$, and let $\mu(x,y)$ be a positive measure in $[0,1] \times L$ with bounded Newtonian potential. We will show that the projection of μ onto L has bounded logarithmic potential. First notice that by calculus

$$\log \frac{1}{|y - t|} + \log\left|\sqrt{1 + (y - t)^2} + 1\right| = \int_x^{x+1} \frac{ds}{\sqrt{(x - s)^2 + (y - t)^2}} .$$

Hence

$$\int \log \frac{1}{|y - t|} \, d\mu(x,y) \leq \int_E \int_x^{x+1} \frac{ds}{\sqrt{(x - s)^2 + (y - t)^2}} \, d\mu(x,y)$$

$$= \int_0^2 ds \int_{\{s-1<x<s\}} \left(\frac{d\mu(x,y)}{\sqrt{(x - s)^2 + (y - t)^2}} \right)$$

$$\leq 2\|U_\mu\|_\infty ,$$

and the projected measure has bounded potential on its support. This means that L has positive logarithmic capacity.

Conversely, if a measure ν on L has bounded logarithmic potential, then for $u + iv \in [0,1] \times L$

$$\int_L \int_0^1 \frac{1}{|u + iv - x - iy|} \, dx d\nu(y)$$

$$= \int_L \left(\log \frac{1}{|y - v|} + \log\left|\sqrt{(y - v)^2 + (u - 1)^2} + (1 - u)\right| \right.$$

$$\left. - \log\left|\sqrt{(y - v)^2 + u^2} - u\right| \right) d\nu(y) \leq \int_L \left(6 + 3 \log \frac{1}{|y - v|} \right) d\nu(y) ,$$

because the second log term is at most 3 while the third is bounded by twice the first plus 3. Therefore the product measure $dxd\nu(y)$

has bounded Newtonian potential.

Exercise 1.5: Show there exists a set L of the type used in the last part of the proof of 1.3.

§2. Cantor Sets

Theorem 1.3 shows that the metric conditions sufficient for $\alpha(E) > 0$ given in §4 of III are not necessary conditions. In this section we show that nevertheless these conditions cannot be improved upon. A special case of the construction given below shows that $\Lambda_1(E) > 0$ does not imply $\gamma(E) > 0$, and hence that the converse of Painlevé's theorem fails.

Let $\lambda = \{\lambda_n\}$ be a non-increasing sequence of real numbers with $\lambda_1 < 1/2$. Form a linear Cantor set $K(\lambda)$ as follows. $K_0 = [0,1]$, $K_1 = [0,\lambda_1] \cup [1 - \lambda_1, 1]$, and at each stage n, K_n is obtained from K_{n-1} by replacing each component of K_{n-1} by its two endmost intervals of length λ_n times the length of the component. Thus $K(\lambda) = \bigcap K_n$ where $K_{n+1} \subset K_n$, and K_n consists of 2^n closed intervals of length

$$\sigma_n = \prod_{j \leq n} \lambda_j .$$

Write $E = E(\lambda) = K(\lambda) \times K(\lambda)$. Then $E = \bigcap_{n \geq 0} E_n$ where E_n is a union of 4^n squares of side σ_n. Denote these squares by $E_{n,j}$ $1 \leq j \leq 4^n$. Let μ be the measure in E such that $\mu(E_{n,j}) = 4^{-n}$ for all n, j.

<u>Lemma 2.1</u>: Let h be a measure function. Then the following are equivalent

(i) $M_h(E(\lambda)) > 0$

(ii) $\underline{\lim}_{n\to\infty} 4^n h(\sigma_n) > 0$

(iii) μ is of growth $h(t)$.

<u>Proof</u>: By 1.5 of III, (iii) implies (i). Since $M_h(E(\lambda)) \leq \underline{\lim}\, 4^n h(\sigma_n)$, (i) implies (ii). Now assume (ii). Let S be a square of side δ and take n such that $\sigma_{n+1} \leq \delta < \sigma_n$. Then S can meet at most 4 squares $E_{n,j}$ so that $\mu(S) \leq 4^{1-n}$. Therefore by (ii) there is a constant c such that $\mu(S) \leq ch(\delta)$.

<u>Theorem 2.2</u>: The following are equivalent

(i) $\Sigma\, 4^{-n}/\sigma_n < \infty$

(ii) There is a measure function h such that $\int \frac{h(t)dr}{t^2} < \infty$, and $M_h(F) > 0$.

(iii) The measure μ on $E = E(\lambda)$ defined by $\mu(E_{n,j}) = 4^{-n}$ for all n, j, has continuous Newtonian potential.

(iv) $C(E) > 0$

(v) $\alpha(E) > 0$

(vi) $\gamma(E) > 0$.

Similar results for linear sets and logarithmic capacity can be found in [62]. Condition (i) was introduced by Denjoy [21], who had proved (i) $\Rightarrow$ (v) by 1931. The equivalence of (i) and (iv) in a more general setting is due to Ohtsuka. See [14], p. 31. The new part here is that (vi) implies (i).

<u>Proof</u>: If (i) holds and $h(\sigma_n) = 4^{-n}$, then

$$\int_0^1 \frac{h(t)}{t^2}\,dt \le \sum_{n=1}^{\infty} \int_{\sigma_{n+1}}^{\sigma_n} \frac{h(t)}{t^2}\,dt \le \sum_{n=1}^{\infty} 4^{-n}(1 - \lambda_1)/\sigma_n .$$

and (ii) holds. If (ii) holds then by Lemma 2.1, μ has growth $h(t)$ so that U_μ is continuous by 4.1 of III. Clearly (iii) $\Rightarrow$ (iv) $\Rightarrow$ (v) $\Rightarrow$ (vi). We now assume that (i) fails, i.e. that

$$\sum 4^{-n}/\sigma_n = \infty$$

and prove $\gamma(E) = 0$. Assume $\gamma(E) > 0$. Let $\gamma_{n,j}$ be a cycle with winding number 1 about $E_{n,j}$ and 0 about $E \setminus E_{n,j}$. For $f \in A(E,1)$ write

$$f_{n,j}(z) = \frac{-1}{2\pi i} \int_{\gamma_{n,j}} \frac{f(\zeta)d\zeta}{\zeta - z}$$

where $z \notin E_{n,j}$ and $\gamma_{n,j}$ is chosen so that its winding number about z is zero. Also write

$$a_{n,j} = a_{n,j}(f) = (f_{n,j})'(\infty) = \frac{1}{2\pi i} \int_{\gamma_{n,j}} f(\zeta)d\zeta .$$

Then $\Sigma_j f_{n,j} = f$ and $\Sigma\, a_{n,j} = f'(\infty)$. While we will not use this fact in the proof, it is useful to notice that the numbers $\{a_{n,j}\}$ determine f, because the projections of E have zero length. We will show that the $a_{n,j}$ are all zero by first considering the case in which $a_{n,j} = 4^{-n} f'(\infty)$ (this amounts to showing that $\hat{\mu}$ is unbounded, if μ is the measure in (iii)) and then reducing the problem to that case.

The proof is broken up into four lemmas. Constants depending only on λ_1 will be denoted by $C_1, C_2, \ldots$.

Lemma 2.3: There is a constant C_1 such that

(a) $f_{n,j} \in A(E_{n,j}, C_1)$

(b) $|a_{n_j}| \leq C_1 \sigma_n$.

Proof: Let $V_{n,j}$ be a square concentric with $E_{n,j}$ but of side $\eta\sigma_n$ where $\eta > 1$ such that the $V_{n,j}$ are pairwise disjoint. For z near $E_{n,j}$ Cauchy's theorem gives

$$f(z) = \frac{1}{2\pi i} \int_{\partial V_{n,j}} \frac{f(\zeta)d\zeta}{\zeta - z} + f_{n,j}(z)$$

so that $|f_{n,j}(z)| \leq C$, where C depends only on η. This implies (a), and (a) implies (b).

Lemma 2.4: Let $z_{n,k}$ be the upper right corner of $E_{n,k}$. Indexing the $E_{n,k}$ so that $0 \in E_{n,1}$, we have

$$\left|\sum_{k=2}^{4^n} \frac{a_{n,k}}{z_{n,k}}\right| \leq C_2 \quad \text{for all } n .$$

Proof: Since $\Sigma_{k=2}^{4^n} f_{n,k}(0) = f(0) - f_{n,1}(0)$ is bounded, we only have to bound $\Sigma_{k=2}^{4^n} g_{n,k}(0)$, where $g_{n,k}(z) = f_{n,k}(z) + a_{n,k}/(z_{n,k} - z)$. Now $g_{n,k}$ is analytic off $\Delta(z_{n,k}, \sqrt{2}\,\sigma_n)$, bounded by C_3, and $g_{n,k}$ vanishes twice at ∞. Two applications of Schwarz's lemma (Chapter I, §2) then give

$$|g_{n,k}(0)| \leq \frac{C_4\sigma_n^2}{(\mathrm{dist}(0,E_{n,k}))^2} .$$

Let

$$B_n = \sigma_n^2 \sum_{k=2}^{4^n} \frac{1}{(\mathrm{dist}(0,E_{n,k}))^2} .$$

Then comparing each $E_{n+1,k}$ to its containing $E_{n,k}$ we have

$$B_{n+1} \leq \frac{3\lambda_{n+1}^2}{(1 - 2\lambda_{n+1})^2} + 4\lambda_{n+1}^2 B_n$$

$$\leq \frac{3\lambda_1^2}{(1 - 2\lambda_1)^2} + 4\lambda_1^2 B_n .$$

Therefore

$$B_n \leq \left(\frac{3\lambda_1^2}{(1 - 2\lambda_1)^2} + B_1 \right) \sum_{j=2}^{n} (4\lambda_1^2)^j .$$

Lemma 2.5: If $a_{n,j} = 4^{-n}f'(\infty)$, then $f'(\infty) = 0$.

Proof: Assuming $f'(\infty) \neq 0$, we have that $4^{-n}\left|\Sigma_{k=2}^{4^n} 1/z_{n,k}\right|$ is bounded for all n. Since all $z_{n,k}$ lie on one quadrant, this means

$$A_n = 4^{-n} \sum_{k=2}^{4^n} \frac{1}{|z_{n,k}|}$$

is bounded. However it is easily seen that

$$A_{n+1} \geq A_n + \frac{2 \cdot 4^{-n}}{\sigma_n}$$

so that because $\Sigma\, 4^{-n}/\sigma_n = \infty$, the A_n are unbounded.

Now let Q be the set of non-increasing sequences $\mu = \{\mu_n\}_{n=1}^{\infty}$ such that $\mu_n \leq \lambda_n$. Then Q is compact in the product topology and the (set valued) mapping $Q \ni \mu \mapsto E(\mu)$ is continuous when we use the Hausdorff metric: $d(K,L) < \varepsilon$ if each lies in an ε neighborhood of the other.

Lemma 2.6: Let $M > 0$, $\gamma > 0$. Then there exists $\delta > 0$ and $N > 0$ such that whenever $\mu \in Q$, $f \in A(E(\mu),M)$ and $f'(\infty) \geq \gamma$ we have

$$\max_{\substack{n \leq N \\ j \leq 4^n}} |a_{n,j}(f)| \geq 4^{-n}(1+\delta)f'(\infty) .$$

Proof: If the assertion is false then there are sequences $\delta_k \searrow 0$, $\mu^{(k)} \in Q$ and $f_k \in A(E(\mu^{(k)}),M)$ such that $f_k'(\infty) \geq \gamma$ and

$$\max_{\substack{n \leq k \\ j \leq 4^n}} |a_{n,j}(f_k)| \leq 4^{-n}(1+\delta_n)f_k'(\infty) .$$

Taking a subsequence we may assume that $\mu^{(k)} \to \mu \in Q$ and the f_k converge uniformly on compacta to a function f. Then $f \in A(E(\mu),M)$ and $f'(\infty) \geq \gamma$. Also

$$a_{n,j}(f) = \lim_{k\to\infty} a_{n,j}(f_k)$$

so that $|a_{n,j}(f)| \leq 4^{-n}f'(\infty)$ for all n and j. Since $\Sigma_{j=1}^{4^n} a_{n,j}(f) = f'(\infty)$, this means that $a_{n,j}(f) = 4^{-n}f'(\infty)$. But Lemma 2.5 holds for $E(\mu)$, so this is impossible.

Conclusion of proof of Theorem 2.2: Assume $f \in A(E(\lambda),1)$ with $f'(\infty) = \gamma > 0$. Let $M \geq \|f_{n,j}\|$ for all n and j. Since $\{\lambda_n, \lambda_{n+1}, \ldots\} \in Q$, Lemma 2.5 applies for each $f_{n,j}$ (after a suitable linear change of variable). Thus

$$\max_j |a_{n,j}| \geq (1 + \delta)4^{-n}\gamma \quad \text{for all} \quad n \geq N ,$$

and repeating this p times with some $f_{N,j}$ we have

$$\max_j |a_{n,j}| \geq (1 + \delta)^p 4^{-n}\gamma \quad \text{for} \quad n \geq pN .$$

Combining this with 2.3(b) we get

$$(1 + \delta)^p 4^{-n}\gamma \leq c_1\sigma_n \quad \text{for} \quad pN \leq n < (p + 1)N .$$

But this means

$$\sum_{\gamma=N}^{\infty} 4^{-n}/\sigma_n \leq c_1/\gamma \sum_{p=1}^{\infty} N(1 + \delta)^{-p} < \infty .$$

If $\lambda_n = \lambda < 1/2$, then $K(\lambda)$ is the Cantor set obtained by removing middle $(1 - 2\lambda)$ths at each step. Then $E_n(\lambda)$ consists of 4^n squares of side λ^n. From 2.1 it follows that $0 < \Lambda_h(E(\lambda)) < \infty$ for $h(t) = t^{-\log 4/\log \lambda}$. Thus $\Lambda_1(E) = 0$ if $\lambda < 1/4$, and $C(E) > 0$ if $\lambda > 1/4$. When $\lambda = 1/4$ we get the example in [34] of a set E with $0 < \Lambda_1(E) < \infty$ but $\gamma(E) = 0$. An earlier example was given in [83]. If the sequence $\{\lambda_n\}$ is chosen so that $4^{-n}/\sigma_n$

tends to zero but $\Sigma\, 4^{-n}/\sigma_n = \infty$, we get a set which has zero analytic capacity but which is not a countable union of sets of finite length.

Now let $h(t)$ be a measure function such that $\int \frac{h(t)}{t^2}\, dr = \infty$. We want to give an example of a set E with $M_h(E) > 0$ but $\gamma(E) = 0$. To apply 2.2 we must make an additional assumption on h: $\log h$ is to be an increasing convex function of $\log t$. Recall that we may assume that h is C^∞ on $(0,\infty)$. Our hypothesis, then, is that

$$a(t) = t\,\frac{d \log h(t)}{dt}$$

is increasing.

Theorem 2.7: Let $h(t)$ be a measure function such that

(i) $\int \frac{h(t)}{t^2}\, dt = \infty$

(ii) $a(t) = \frac{th'(t)}{h(t)}$ is increasing.

Then there is a Cantor set $E = E(\mu)$ such that $M_h(E) > 0$ but $\gamma(E) = 0$.

Proof: Let $\sigma_n = h^{-1}(4^{-n})$. Then

$$\log 4 = \int_{\sigma_{n+1}}^{\sigma_n} d \log h = \int_{\sigma_{n+1}}^{\sigma_n} \frac{a(t)}{t}\, dt = a(t_n) \log \frac{\sigma_n}{\sigma_{n+1}}$$

with $\sigma_{n+1} \le t_n \le \sigma_n$. Thus $\frac{\sigma_n}{\sigma_{n+1}}$ increases, so λ_n decreases. Replacing σ_n by $c\sigma_n$ for c small we can assume $\sigma_1 = \lambda_1 < 1/2$. Then the sequence $\{\lambda_n = \sigma_n/\sigma_{n-1}\}$ determines a Cantor set $E = E(\lambda)$ such that $M_h(E) > 0$. Since $\int_{\sigma_{n+1}}^{\sigma_n} \frac{h(t)}{t^2}\, dt \le 4^{-n}/\sigma_n$ and $\int \frac{h(t)}{t^2}\, dt = \infty$, $\gamma(E) = 0$ by Theorem 2.2.

Examples of measure functions satisfying 2.7 (i) and (ii) include

$$h(t) = \frac{t}{(\log 1/t)(\log\log 1/t)} .$$

Condition 2.7(ii) is needed because the proof of 2.2 required that $\lambda_{n+1} \leq \lambda_n$.

Problem: Let $\lambda_n \leq a < 1/2$, and let $E = E(\lambda)$ be the corresponding Cantor set. Prove that $\gamma(E) > 0$ if and only if $\Sigma\, 4^{-n}/\sigma_n < \infty$, where $\sigma_n = \Pi_{j=1}^n \lambda_j$.

§3. Vitushkin's Example

Let $h(t)$ be a measure function such that $h(t)/t$ tends to 0 as t tends to 0. We now give the construction from [84] of a compact set E and a function f such that

1°. $M_h(E) = 0$

2°. E is totally disconnected

3°. $f \in C(E,M)$ for some M, $f \neq 0$.

4°. If γ is a Jordan curve disjoint from E then

$$\int_\gamma f(z)dz = 0 .$$

5°. There is no measure μ on E such that $f(z) = \hat{\mu}(z)$ for all $z \notin E$.

Notice 5° follows from 4°, because the function f has Pompeiu variation 0 and consequently can not be represented as a Cauchy transform. One can also modify the construction so that $\int_\gamma z^k f(z)dz = 0$ for all k in some finite set, and we outline the modified construction. Originally the construction was done so that

$$\sup \sum_{j=1}^{n} \left| \int_{\gamma_j} zf(z)dz \right| = \infty ,$$

where the supremum is taken over all collections $\gamma_1,\ldots,\gamma_n$ of pairwise disjoint rectifiable curves whose union surrounds E. This means that f cannot be represented as a "Golubev series"

$$f(z) = \sum_{k=1}^{\infty} \int \frac{d\mu_k(\zeta)}{(\zeta - z)^k}$$

where $\{\mu_k\}$ is a sequence of measures in E. We will ignore this aspect of the problem; for if the reader is referred to §3 of [84].

Before giving this rather complicated construction we digress momentarily to motivate it. The set E will be a decreasing intersection $\bigcap E_n$, where each E_n consists of a union of squares, each having some "antennae" (see Figure 1). The purpose of the antennae is simply this: if γ is the boundary of some square crossing E, then γ cannot be approximated by a sequence of curves which are disjoint from E and which have bounded lengths. This is why 3° and 4° can hold simultaneously.

<u>Construction</u>: Define inductively a sequence $\{E_n\}$ of sets and a sequence $\{\lambda_n(z)\}$ of functions such that

(a) For any $\varepsilon > 0$, E_n can be expressed as a union of squares $k_1, k_2, \ldots, k_q$, $q = q(n,\varepsilon)$ with their sides parallel to their axes, their interiors pairwise disjoint, and their sides all of the same length, which is less than ε. These are to be indexed so that for any component K of E_n, $\{j : k_j \subset K\}$ is an interval and so that if k_j, k_{j+1} are contained in K then k_j and k_{j+1} have a common side.

(b) $E_{n+1} \subset E_n$ and the maximum diameter of the components of E_n tend to 0 as $n \to \infty$.

(c) $\lambda_n(z) = 0$ for $z \notin E_n$, λ_n is constant on each k_j; but $\lambda_n \neq 0$.

(d) If K is a component of E_n then $\iint_K \lambda_n dxdy = 0$.

To begin take E_1 the unit square $[0,1] \times [0,1]$ and let $\lambda_1 = 1$ on the left half of E_1, and $\lambda_1 = -1$ on the right half of E_1.

Assume $E_1, E_2, \ldots, E_{n-1}$ and $\lambda_1, \ldots, \lambda_{n-1}$ have been constructed satisfying (a)-(d). Express E_{n-1} as a union of squares $k_1, k_2, \ldots, k_q$, $q = q_n$ satisfying the condition (a). Let k_j' be a square concentric with k_j but having side $r_n = \text{side}(k_j)/m_n$. Fix k_j and k_{j+1} with a common side. Assume for convenience that k_j lies to the left of k_{j+1}. Join their centers by a line segment and let $a_j \in k_j'$, $b_j \in k_{j+1}'$ be the points of intersection of this line segment with $\partial k_j'$ and $\partial k_{j+1}'$ respectively. Let $0 < \delta = \delta_n < r_n/2$ and let $s = s_n$ be a positive integer. We take some closed rectangles (the "antennae") α_j^ℓ and β_{j+1}^ℓ, $1 \leq \ell \leq \delta_n$ as follows:

> α_j^ℓ has lower left vertex $a_j + i\delta(\ell - 1)/s$,
> base $(b_j - a_j)/2 + \delta/2$, and altitude $\delta/4s$.
>
> β_{j+1}^ℓ has lower right vertex $b_j + i\delta(\ell - 1)/s + i\delta/2s$,
> and the same dimensions as α_j^ℓ.

If k_j is not to the left of k_{j+1} but $k_j \cap k_{j+1} \neq \emptyset$, construct the α_j^ℓ and β_{j+1}^ℓ as above after rotation yourself so that k_j is to the left of k_{j+1}.

Set $\alpha_j = \bigcup_\ell \alpha_j^\ell$, $\beta_{j+1} = \bigcup_\ell \beta_{j+1}^\ell$. If k_j and k_{j+1} lie in different components of E_{n-1}, take $\alpha_j = \beta_{j+1} = \emptyset$. Now write

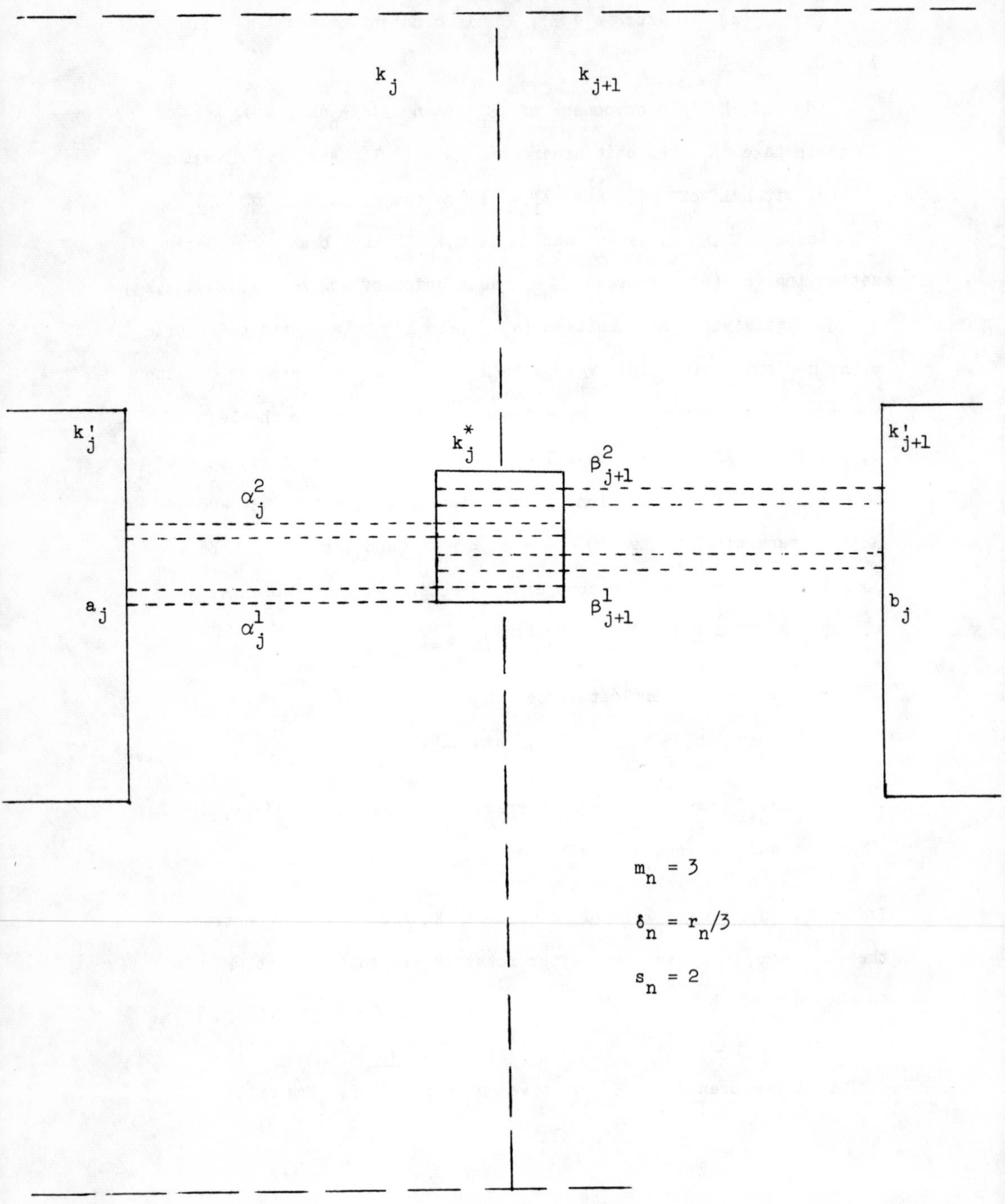

k_j
k_{j+1}
k'_j
k^*_j
β^2_{j+1}
k'_{j+1}
α^2_j
a_j
α^1_j
β^1_{j+1}
b_j
$m_n = 3$
$\delta_n = r_n/3$
$s_n = 2$

Figure 1

$$E_n = \bigcup_j k'_j \cup \bigcup_j (\alpha_j \cup \beta_{j+1}) .$$

The components of E_n are $\beta_j \cup k'_j \cup \alpha_j$, and if q_n is sufficiently big, that is, if the k_j are sufficiently small, the maximum of the diameters of these compoents goes to zero. Hence (b) holds. Also, if δ_n/r_n is a rational number, it is not hard to see that E_n satisfies condition (a).

We now define λ_n. Set $\lambda_n(z) = 0$ for $z \notin E_n$. On k'_j take $\lambda_n(z) = m_n^2\lambda_{n,j}$ where $\lambda_{n,j}$ is the constant value of $\lambda_{n-1}(z)$ on k_j. Let k_j^* be the square with center $(a_j + b_j + i\delta)/2$ and side $\delta = \delta_n$. Declare $\lambda_n(z) = 0$ if $z \notin \bigcup(k'_j \cup k_j^*)$. Finally, on the sets $\alpha_j \cap k_j^*$, $\beta_{j+1} \cap k_j^*$, $\lambda_n(z)$ is defined inductively as follows:
For $z \in \alpha_1 \cap k_1^*$, define $\lambda_n(z) = -4m^2\lambda_{n,1} \cdot r_n^2/\delta_n^2$.
For $z \in \beta_2 \cap k_1^*$, define $\lambda_n(z) = 4m^2\lambda_{n,1} \cdot r_n^2/\delta_n^2$.
If λ_n has been defined on $\beta_j \cap k_{j-1}^*$, then for $z \in \alpha_j \cap k_j^*$, define

$$\lambda_n(z) = -4/\delta^2 \int_{\beta_j \cup k'_j} \lambda_n(\zeta)d\xi d\eta ,$$

and for $z \in \beta_{j+1} \cap k_j^*$, define

$$\lambda_n(z) = 4/\delta^2 \int_{\beta_j \cup k'_j} \lambda_n(\zeta)d\xi d\eta .$$

Then (c) holds. It follows that for every j for which $\alpha_j \neq \emptyset$,

$$\int_{\alpha_j \cup k'_j \cup \beta_j} \lambda_n(z)dxdy = 0 .$$

And for each component K of E_{n-1} there is only one index p (the largest, in fact) such that $k_p \subset K$ and $\alpha_p = \emptyset$. Then by the above

$$\int_{k_p' \cup \beta_p} \lambda_n(z)dxdy = \sum_{j=j_0}^{p} \int_{k_j'} \lambda_n(z)dxdy$$

$$+ \sum_{j=j_0}^{p} \int_{\alpha_j \cup \beta_{j+1}} \lambda_n(z)dxdy ,$$

where j_0 is the first index such that $k_{j_0} \subset K$. The second sum is zero, while the first is

$$\sum_{j=j_0}^{p} \int_{k_j} \lambda_{n-1}(z)dxdy = \int_K \lambda_{n-1}(z)dxdy = 0 .$$

Consequently (d) holds.

Let $E = \bigcap E_n$ and $f(z) = \lim_{n\to\infty} f_n(z)$ where

$$f_n(z) = \int_{E_n} \frac{\lambda_n(\zeta)}{\zeta - z} d\xi d\eta .$$

In order to verify 1°-4° we must select the parameters q_n, m_n, δ_n and s_n suitably.

Lemma 3.1: If $q = q_n$ is fixed, then

$$\lim_{s_n\to\infty} \left\| \int_{\alpha\cup\beta} \frac{\lambda_n(\zeta)}{\zeta - z} d\xi d\eta \right\|_\infty = 0$$

where

$$\alpha \cup \beta = \bigcup_{j=1}^{q-1} \alpha_j \cup \beta_{j+1} .$$

Proof: It suffices to show

$$\left\| \int_{\alpha_j \cup \beta_{j+1}} \frac{\lambda_n(\zeta)}{\zeta - z}\, d\xi d\eta \right\|_\infty \to 0$$

for every j. Let

$$g_s(z) = \int_{\alpha_j} \frac{\lambda_n(\zeta)}{\zeta - z}\, d\xi d\eta \ .$$

Since $\|\lambda_n\|_\infty$ is bounded independent of s, the functions $g_s(z)$ are equicontinuous (see III §4). Hence

$$\int_{\alpha_j \cup \beta_{j+1}} \frac{\lambda_n(\zeta)}{\zeta - z}\, d\xi d\eta = g_s(z) - g_s(z + \delta/4s)$$

tends to 0 uniformly on z as s becomes large.

<u>Lemma 3.2</u>: Let

$$m_n = r_n^{-3/4}(h(r_n))^{1/4} \ .$$

Then

$$\lim_{q\to\infty} \left\| f_{n-1}(z) - \int_{k'} \frac{\lambda_n(\zeta)}{\zeta - z}\, d\xi d\eta \right\|_\infty = 0$$

where

$$k' = \bigcup_{j=1}^{q-1} k_j' \ .$$

Notice that the hypothesis of 3.2 does not fix r_n but merely states that the squares k_j are to have side $m_n r_n = (r_n h(r_n))^{1/4}$. Taking q large only amounts to taking this number small, which can be done by (a).

Proof of 3.2: For any q

$$f_{n-1}(z) - \int_{k'} \frac{\lambda_n(\zeta)}{\zeta - z}\, d\xi d\eta = \sum_{j=1}^{q} g_j(z)$$

where

$$g_j(z) = \int_{k_j} \frac{\lambda_{n-1}(\zeta) d\xi d\eta}{\zeta - z} - m_n^2 \int_{k_j'} \frac{\lambda_{n-1}(\zeta) d\xi d\eta}{\zeta - z} .$$

Now $\|\lambda_{n-1}\|_\infty$ is independent of q, and we are going to show $\Sigma_{j=1}^{q} |g_j(z)| \to 0$ $(q \to \infty)$ so we will assume $\|\lambda_{n-1}\| = 1$. $c_1, c_2, \ldots$ below will denote constants. Now $|g_j(z)| \leq c_1 m_n r_n$, g_j is analytic off k_j, and $g_j(\infty) = g_j'(\infty) = 0$. By Schwarz's lemma we have

$$|g_j(z)| \leq c_2 (m_n r_n)^3 / (\mathrm{dist}(z, k_j))^2 .$$

Now for $p = 1, 2, \ldots$

$$p r_n m_n \leq d(z, k_j) < (p+1) r_n m_n$$

for at most $c_3 p$ values of j, when $p \leq 1 + 2/r_n m_n$, and for no values of j otherwise. Therefore,

$$\sum |g_j(z)| \leq c_1 m_n r_n + c_4 \sum_{p=1}^{1+2/r_n m_n} \frac{p m_n r_n}{p^2}$$

$$\leq m_n r_n (c_1 + c_5 \log(1 + 2/r_n m_n)) .$$

Since $m_n r_n \to 0$ as $q \to \infty$, we have

$$\lim_{q\to\infty} \left\| f_{n-1}(z) - \int_{k'} \frac{\lambda_n(\zeta)}{\zeta - z}\, d\xi d\eta \right\| \leq \|\lambda_{n-1}\| \lim_{m_n r_n \to 0} \sum_{j=1}^{q} |g_j(z)| = 0 \, .$$

Lemmas 3.1 and 3.2 imply that if we declare $m_n r_n = (r_n h(r_n))^{1/4}$, take q_n sufficiently large and then take s_n sufficiently large, we have $\|f_n - f_{n-1}\| < 3^{-n}$. Then f_n converges uniformly to a function f, $f \neq 0$ and 2°, 3° and 4° hold. The next lemma shows that if q_n is sufficiently large and if the remaining parameter δ_n is sufficiently small, then $M_h(E) = 0$.

<u>Lemma 3.3</u>: Let $m_n = r_n^{-3/4} h(r_n)^{1/4}$. If δ_n is sufficiently small then E_n can be covered by discs of radii $\rho_1, \rho_2, \ldots, \rho_\nu$ such that

$$\sum h(\rho_i) \leq (h(r_n)/r_n)^{1/2} \, .$$

Consequently $M_h(E) \leq \lim (h(r_n)/r_n)^{1/2} = 0$.

<u>Proof</u>: Recall $E_n = \bigcup_{j=1}^{q_n} k_j' \cup \bigcup_{j=1}^{q_n} (\alpha_j \cup \beta_j)$. Each k_j' can be covered by a disc of radius r_n. Take $\rho_i = r_n$ for $i \leq q_n$. Each set $\alpha_j \cup \beta_{j+1}$ lies in a rectangle with height δ_n and base no larger than $m_n r_n$. It can be covered by $(m_n r_n/\delta_n) + 1$ discs of radius δ_n. Take $\rho_i = \delta_n$ for $i > q_n$. At most $q_n + q_n m_n r_n/\delta_n$ such ρ_i need be used. With this covering we have

$$\sum h(\rho_i) \leq q_n h(r_n) + q_n h(\delta_n) + q_n m_n r_n h(\delta_n)/\delta_n \, .$$

Now $q_n h(r_n) = q_n m_n^4 r_n^3$, and $q_n m_n^2 r_n^2 = \text{Area}(E_{n-1})$. Therefore

$$q_n h(r_n) \leq \text{Area}(E_{n-1}) m_n^2 r_n = \text{Area}(E_{n-1})(h(r_n)/r_n)^{1/2} \, .$$

So if $h(\delta_n)$ and $h(\delta_n)/\delta_n$ are sufficiently small, we have

$$\sum h(\rho_i) \leq (h(r_n)/r_n)^{1/2} .$$

We now describe how to modify the construction so that whenever γ is disjoint from E

$$\int_\gamma f(z)z^k dz = 0$$

for $k \leq N$, where N is some fixed integer. The set E is the same, but the functions $\lambda_n(\zeta)$ are to be altered so that for all j,

$$\int_{k_j' \cup \alpha_j \cup \beta_j} \lambda_n(\zeta)\zeta^k d\xi d\eta = 0 \quad \text{for } k \leq N .$$

Let $\lambda_1(\zeta)$ be any bounded function so that $\int_{E_1} \lambda_1(\zeta)\zeta^k d\xi d\eta = 0$ for all $k \leq N$, and $\lambda_1(\zeta) = 0$ off E_1, $\lambda_1(\zeta) \neq 0$ on E_1. Assuming λ_{n-1} has been defined so that λ_{n-1} is in $L^\infty(dxdy)$ and λ_{n-1} is orthogonal to ζ^k, $k \leq N$, over each component of E_{n-1}. Again declare $\lambda_n(z) = 0$ if $z \notin \bigcup(k_j' \cup k_j^*)$. Let w_j be the center of k_j and define $\lambda_n(z)$ for $z \in k_j'$ by $\lambda_n(z) = m_n^2 \cdot \lambda_{n-1}(m_n(z - w_j))$. A glance at its proof shows that 3.2 still holds with this choice of λ_n. Now fix q_n so that we can use 3.2 and then fix δ_n so that we can use 3.3. In order to define λ_n on the sets $\alpha_j \cap k_j^*$ and $\beta_{j+1} \cap k_j^*$ we need the following lemma.

<u>Lemma 3.4</u>: Let $\eta > 0$ and let F be a compact subset with area$(F) = \eta$. Let $\tau_0, \tau_1, \ldots, \tau_N$ be complex numbers with $|\tau_j| \leq 1$ for all j. Then

there is a constant $M = M(\eta, N)$ and a function $\lambda \in L^\infty(K)$ such that $\|\lambda\|_\infty \le M$ and

$$\int_K \lambda(\zeta)\zeta^k d\xi d\eta = \tau_k \qquad 0 \le k \le N .$$

<u>Proof</u>: Let $g(z) = \iint_F \frac{d\xi d\eta}{z - \zeta}$. Then $\|g\|_\infty \le c_1\eta$ and

$$g(z) = a_1/z + a_2/z^2 + \cdots + a_N/z^N + \cdots$$

where $a_k = \iint_F \zeta^k d\xi d\eta$, so that $|a_k| \le c_2\eta$, and $a_1 = \eta$. Let $G_N = g^N/a_1^N$, $G_{N-1} = g^{N-1}/a_1^{N-1} - Na_2a_1^{N-1}G_N$, and so forth, so that for $0 \le j \le N$,

$$G_j(z) = 1/z^j + b_j/z^{N+1} + \cdots .$$

As suggested above this can be done by taking $G_j(z) = p_j(g(z))$ where p_j is a polynomial of degree at most N and the coefficients of p_j do not exceed $c_3(\eta, N)$. Now

$$G_j(z) = \iint_F \frac{\lambda_j(\zeta)}{z - \zeta} d\xi d\eta$$

where $\lambda_j = \pi \frac{\partial G_j}{\partial \bar{z}}$. But $\frac{\partial G_j}{\partial \bar{z}} = p_j'(g(z)) \frac{\partial g}{\partial \bar{z}} = \pi p_j'(1)$ for $z \in F$, so that $|\lambda_j| \le c_4(\eta, N)$. Since

$$\int \zeta^k \lambda_j(\zeta) d\xi d\eta = \delta_{jk}, \quad j,k \le N$$

the function $\lambda = \Sigma\, \tau_k \lambda_k$ has the desired properties.

Returning to the definition of λ_n, we assume λ_n has been defined on $\beta_j \cap k_{j-1}^*$ and then define λ_n on $\alpha_j \cap k^*$ so that

$$\int_{\alpha_j} \lambda_n(\zeta)\zeta^k d\xi d\eta = - \int_{k_j^! \cup \beta_j} \lambda_n(\zeta)\zeta^k d\xi d\eta$$

for all $k \leq N$ and so that $|\lambda_n(\zeta)| \leq M_1$ on α_j where M_1 depends on the bounds for values of λ_n defined previously, but not on the parameter s_n, because the area of α_j is independent of s_n. On $\beta_{j+1} \cap k_j^*$ define $\lambda_n(\zeta) = \lambda_n(\zeta - i\delta/2s) + \mu_n(\zeta)$ where $\mu_n(\zeta)$ is chosen so that for $k \leq N$,

$$\int_{\beta_{j+1} \cap k_j^*} \zeta^k \mu_n(\zeta) d\xi d\eta = \int_{\alpha_j} (\zeta^k - (\zeta + i\delta/2s)^k)\lambda_n(\zeta) d\xi d\eta = \varepsilon_k,$$

and such that $\|\mu_n\|_\infty$ is minimal. It follows that

$$\int_{\alpha_j \cup k_j \cup \beta_j} \zeta^k \lambda_n(\zeta) d\xi d\eta = 0$$

for all j. Finally, we must verify Lemma 3.1 in this case. Since $|\lambda_n| \leq M_1$ on α_j, with M_1 independent of s_n, we can make each $|\varepsilon_k|$ as small as we please by taking s_n sufficiently large. By Lemma 3.4 we can then take $\|\mu_n\|_\infty$ small, so that

$$\left\| \int_{\beta_j} \frac{\mu_n(\zeta)}{\zeta - z} d\xi d\eta \right\| \leq c_1 \|\mu_n\|_\infty \delta$$

is as small as we please. Again since M_1 is independent of s_n,

$$\left\| \int_{\alpha_j} \frac{\lambda_n(\zeta)}{\zeta - z} d\xi d\eta + \int_{\beta_j} \frac{\lambda_n(\zeta) - \mu_n(\zeta)}{\zeta - z} d\xi d\eta \right\|$$

$$= \left\| \int_{\alpha_j} \lambda_n(\zeta) \left(\frac{1}{\zeta - z} - \frac{1}{\zeta - i\delta/2s - z} \right) d\xi d\eta \right\|$$

tends to 0 as $s_n \to \infty$, just as in the proof of 3.1.

CHAPTER V. APPLICATIONS TO APPROXIMATION

§1. Algebras and Capacities

Let E be a subset of the plane and recall that S_ν denotes the closed support of a compactly supported measure ν. Define $B(E)$ as the uniformly closed linear span of

$$\{1\} \cup \{\hat{\nu} : S_\nu \subset E \text{ and } U_{|\nu|} \text{ is uniformly convergent}\} .$$

Notice that $B(E)$ consists of functions continuous on S^2 and analytic on $S^2 \setminus E$. When $h(t)$ is a measure function satisfying the integrability condition

$$(*) \qquad \int_0 \frac{h(t)}{t^2}\, dt < \infty$$

we also define $B_h(E)$ as the uniformly closed linear span of

$$\{1\} \cup \{\nu : S_\nu \subset E \text{ and } |\nu| \text{ has growth } h(t)\} .$$

Thus $B_h(E) \subset B(E)$, and cases occur where $B(E) \neq C$ while $B_h(E) = C$ for all h. (See the remark following III 4.1.)

<u>Lemma 1.1</u>: $B(E)$ and $B_h(E)$ are algebras.

<u>Proof</u>: Let $f(z) = \hat{\mu}(z)$ and $g(z) = \hat{\nu}(z)$ where μ and ν are measures with compact supports contained in E and absolutely convergent Newtonian potentials. A calculation shows that $fg = \hat{\sigma}$ where

$$\sigma = f\nu + g\mu$$

is a measure on E and $U_{|\sigma|}$ is absolutely convergent. This shows $B(E)$ is an algebra. If $|\nu|$ and $|\mu|$ have growth $Ch(t)$, then so does σ (with a larger constant C), and thus $B_h(E)$ is an algebra.

Roughly speaking, the algebras $B(E)$ and $B_h(E)$ lie somewhere between an algebra $R(K)$ and an algebra $A(\Omega)$. In this chapter we make that more precise by considering two approximation problems. The first problem consists of deciding when $B(E)$ or $B_h(E)$ is contained in $R(K)$ where K is a compact set with $K^0 \cap E = \emptyset$. It is discussed in Section 4. For the second problem let E be a compact set. We then want to know when a function in $C(E,1)$ is in $B(E)$ or $B_h(E)$.

We begin by considering the second problem. It will be solved by defining the appropriate "analytic capacities" and applying, with minor modifications, Vitushkin's theory of rational approximation. However the capacities used for our problem are sufficiently semi-additive to yield results whose analogue for rational approximation remains unresolved. An outstanding question is the following: If $\alpha(F) = 0$, is $C(E\setminus F,1)$ dense in $C(E,1)$ for all compact sets E? Equivalently, we can ask that $\alpha(E) \leq C\alpha(E\setminus F)$ for all compact E whenever $\alpha(F) = 0$. We can answer affirmatively the corresponding question for approximation by Cauchy transforms because the "capacity" used to approximate by absolutely convergent Cauchy transforms has the same null sets as the Newtonian capacity; and for approximation by Cauchy transforms of measures of growth $h(t)$ the "capacity" has the same null sets as the Hausdorff measure Λ_h.

For capacities we define, for E bounded

$$\gamma_C(E) = \sup\{|f'(\infty)| : f \in B(E), \|f\|_\infty \le 1\}$$

$$\gamma_h(E) = \sup\{|f'(\infty)| : f \in B_h(E), \|f\|_\infty \le 1\} .$$

Using 3.3 of III it is easily seen that

$$\gamma_C(E) \ge C(E) ,$$

and $\gamma_C(E) = 0$ if and only if $C(E) = 0$. Also, $\gamma_h(E) = 0$ if $m_h(E) = 0$, and if $\gamma_h(E) = 0$, then $\underline{m_h}(E) = \sup\{m_h(F) : F \text{ compact}, F \subset E\} = 0$. (See §1 of III). Hence if E is an analytic set, $\gamma_h(E) = 0$ if and only if $m_h(E) = 0$. Clearly

$$\gamma_h(E) \le \gamma_C(E) \le \alpha(E) .$$

The hypotheses of approximation theorems will often consist of special cases of the reverse inequalities.

The property of $B(E)$ and $B_h(E)$ whose analogue for rational approximation is not known is the following.

<u>Theorem 1.2</u>: Let L have outer Newtonian capacity zero. Then for any set E, $B(E\backslash L)$ is dense on $B(E)$. Similarly, if $\underline{m_h}(L) = 0$ then for any E, $B_h(E\backslash L)$ is dense in $B_h(E)$.

<u>Proof</u>: Let $f \in B(E)$ and for convenience suppose $f(\infty) = 0$. Let ν be a measure on E such that $U_{|\nu|}$ is uniformly convergent and $\|\hat{\nu} - f\|_\infty$ is small. By hypothesis $|\nu|(L) = 0$, so there is a neighborhood V of L such that for $\varepsilon > 0$

$$\int_V \frac{d|\nu|(\zeta)}{|\zeta - z|} < \varepsilon$$

for all z. Then

$$g = \int_{E \setminus V} \frac{d\nu(\zeta)}{\zeta - z}$$

is in $B(E \setminus L)$ and $\|g - f\|_\infty$ is small. The proof for B_h is the same.

Corollary 1.3: Let L be an analytic set. If $\gamma_C(L) = 0$, then $\gamma_C(E) = \gamma_C(E \setminus L)$, for all E. If $\gamma_h(L) = 0$, then $\gamma_h(E) = \gamma_h(E \setminus L)$, for all E.

§2. Characterizations of $B(E)$ and $B_h(E)$

Fix a measure function h satisfying

$$(*) \qquad \int_0 \frac{h(t)}{t^2}\, dt < \infty .$$

In this section we give two conditions, each necessary and sufficient for a function to be in $B_h(E)$. We also give similar conditions for a function to be in $B(E)$. The conditions are parallel to Vitushkin's conditions that a function lie in the algebra $R(K)$ [81], and the proofs are formally the same as Vitushkin's. For that reason we only prove some preliminary lemmas; the remaining steps can be given by substituting γ_h or γ_C for α in the arguments in the sources [28] or [81]. We begin with the only two lemmas whose proofs depend on the character of $B(E)$ or $B_h(E)$.

Lemma 2.1: Let A be either $B_h(E)$ or $B(E)$. If $f \in A$, then

(i) $z(f(z) - f(\infty)) \in A$

(ii) $\dfrac{f(z) - f(z_0)}{z - z_0} \in A$ if $z_0 \notin E$

(iii) $|f(z_0)| \le \dfrac{2\|f\|\tilde{\gamma}(E)}{\operatorname{dist}(z_0,E)}$ if $z_0 \notin E$

where $\tilde{\gamma} = \gamma_h$ if $A = B_h(E)$ and $\tilde{\gamma} = \gamma_C$ if $A = B(E)$.

<u>Proof</u>: We may assume $f(z) = a + \hat{\nu}(z)$ where $S_{|\nu|} \subset E$. Then

$$z(f(z) - f(\infty)) = \int \left(\frac{1}{\zeta - z} - \frac{1}{\zeta} \right) \zeta d\nu(\zeta)$$

$$= \hat{\mu}(z) - \hat{\mu}(0) ,$$

where $\mu = \zeta\nu$. If $U_{|\nu|}$ is uniformly convergent, then so is $U_{|\mu|}$, and if $|\nu|$ has growth $ch(t)$ then $|\mu|$ has growth $c'h(t)$. Hence $\hat{\mu} \in A$ and (i) holds. Also

$$\frac{f(z) - f(z_0)}{z - z_0} = \frac{\hat{\nu}(z) - \hat{\nu}(z_0)}{z - z_0} = \int \frac{1}{\zeta - z} \frac{1}{\zeta - z_0} d\nu(\zeta)$$

$$= \hat{\sigma}(z) ,$$

where $\sigma = (\zeta - z_0)^{-1}\nu$. Since $z_0 \notin E$, $|\sigma|$ is boundedly absolutely continuous to ν and so $\hat{\sigma} \in A$ just as above. Hence (ii) holds. Now (iii) follows from (ii) as in the proof of the Schwarz lemma, I 2.1.

For $\varphi \in C_0^\infty$ the Vitushkin localization operator is defined by

$$T_\varphi f(z) = \varphi(z)f(z) + \frac{1}{\pi} \iint \frac{f(\zeta)}{\zeta - z} \frac{\partial\varphi}{\partial\bar{\zeta}} d\xi d\eta \qquad \zeta = \xi + i\eta$$

$$= \frac{1}{\pi} \iint \frac{f(\zeta) - f(z)}{\zeta - z} \frac{\partial\varphi}{\partial\bar{\zeta}} d\xi d\eta .$$

See [28] or [81]. The domain of this operator can be all locally integrable functions f, but we will be working with continuous functions. The two formulae defining $T_\varphi f$ coincide becasse of Green's theorem. By Leibnitz's rule and 1.2 of II we have (in the sense of distributions)

$$\frac{\partial}{\partial \bar{z}} T_\varphi f = \frac{\partial \varphi}{\partial \bar{z}} \cdot f + \varphi \frac{\partial f}{\partial \bar{z}} - f \frac{\partial \varphi}{\partial \bar{z}} = \varphi \frac{\partial f}{\partial \bar{z}} .$$

Consequently the operator T_φ localizes the singularities of a function f; that is its important feature.

<u>Lemma 2.2</u>: If $f(z) = \hat{\mu}(z)$, then $T_\varphi f(z) = \widehat{\varphi\mu}(z)$. If $f \in B_h(E)$ or $f \in B(E)$ then $T_\varphi f \in B_h(E \cap \text{Supp } \varphi)$ or $T_\varphi f \in B(E \cap \text{Supp } \varphi)$.

<u>Proof</u>: We showed above that $T_\varphi \hat{\mu} = \widehat{\varphi\mu}$. If $U_{|\mu|}$ is uniformly convergent then so is $U_{|\varphi\mu|}$, and if $|\mu|$ has growth $ch(t)$, $|\varphi\mu|$ has growth $c'h(t)$. Approximating, we see that $B_h(E)$ and $B(E)$ are invariant under the operators T_φ.

<u>Theorem 2.3</u>: Let f be continuous on S^2 and let E be a compact plane set. Then $f \in B_h(E)$ if and only if there exist $r \geq 1$ and $\Omega(\delta)$ tending to 0 with δ such that

$$\left| \iint f \frac{\partial f}{\partial \bar{z}} \, dxdy \right| \leq \Omega(\delta)\delta \left\| \frac{\partial f}{\partial \bar{z}} \right\| \gamma_h(\Delta(z_0, r\delta) \cap E)$$

for all $\varphi \in C_0^\infty$ with support in $\Delta(z_0, \delta)$. Similarly, $f \in B(E)$ if and only if there exist $r \geq 1$ and $\Omega(\delta)$ tending to 0 with δ such that

$$\left|\iint f \frac{\partial\varphi}{\partial\bar{z}}\,dxdy\right| \leq \Omega(\delta)\delta\left\|\frac{\partial\varphi}{\partial\bar{z}}\right\|\gamma_C(\Delta(z_0;\ r\delta)\cap E)$$

for all $\varphi \in C_0^\infty(\Delta(z_0,\delta))$.

Proof: (Outline) One way is easy. If $f \in B_h(E)$, then $T_\varphi f \in B_h(E \cap \Delta(z_0,\delta))$ and

$$\|T_\varphi f\|_\infty \leq \frac{2}{\pi}\,\omega_f(\delta)\left\|\frac{\partial\varphi}{\partial\bar{z}}\right\| \sup_z \iint_{\Delta(z_0,\delta)} \frac{d\xi d\eta}{|\zeta - z|}$$

$$\leq 4\delta\omega_f(\delta)\left\|\frac{\partial\varphi}{\partial\bar{z}}\right\| ,$$

where $\omega_f(\delta)$ is the modulus of continuity. Since

$$\iint f \frac{\partial\varphi}{\partial\bar{z}}\,dxdy = \pi(T_\varphi f)'(\infty) ,$$

we have

$$\left|\iint f \frac{\partial\varphi}{\partial\bar{z}}\,dxdy\right| \leq 4\pi\omega_f(\delta)\delta\left\|\frac{\partial\varphi}{\partial\bar{z}}\right\|\gamma_h(E \cap \Delta(z_0,\delta)) ,$$

The proof for $B(E)$ is the same.

The converse is much harder. It involves letting φ run through a partition of unity and matching two coefficients of each $T_\varphi f$. The reasoning is formally the same as that on pp. 172-177 of [81] or 214-217 of [28] and we refer the reader there for the details.

Theorem 2.4: Let f be continuous on S^2 and let E be a compact plane set. Then $f \in B_h(E)$ if and only if there exists $\Omega(\delta)$ tending to 0 with δ such that for every square S of side δ,

$$\left| \int_{\partial S} f(z)dz \right| \leq \Omega(\delta)\gamma_h(S \cap E) .$$

Similarly, $f \in B(E)$ if and only if there is such a function $\Omega(\delta)$ such that

$$\left| \int_{\partial S} f(z)dz \right| \leq \Omega(\delta)\gamma_C(S \cap E) .$$

<u>Proof</u>: The sufficiency of the conditions follows from Theorem 2.3 using the argument on pp. 177-180 of [81]. The converse is proved using the Melnikov estimates for line integrals but with γ_h or γ_C instead of α. We will never use the converse, and we omit the argument. See Chapter III of [81] or pp. 228-234 of [28].

The proof of the next theorem is similar to that of the analogous result in [28] or [81]. Notice that hard part (c) $\Rightarrow$ (a) follows from 2.3.

<u>Theorem 2.5</u>: Let E be a compact plane set and let $\Omega = S^2\backslash E$. The following three conditions are equivalent.

(a) $A(\Omega) = B_h(E)$

(b) For every disc $\Delta(z_0,\delta)$,

$$\alpha(\Delta(z_0,\delta) \cap E) = \gamma_h(\Delta(z_0,\delta) \cap E) .$$

(c) For each $z_0 \in E$ there are $\delta_0 > 0$, $r_0 \geq 1$, $C > 0$ such that for $\delta < \delta_0$

$$\alpha(\Delta(z_0,\delta) \cap E) \leq C\gamma_h(\Delta(z_0,r\delta) \cap E) .$$

Similarly, the next three conditions are equivalent.

(a') $A(\Omega) = B(E)$

(b') For every disc $\Delta(z_0,\delta)$

$$\alpha(\Delta(z_0,\delta) \cap E) = \gamma_C(\Delta(z_0,\delta) \cap E) .$$

(c') For each $z_0 \in E$ there are $\delta_0 > 0$, $r \geq 1$ and c such that for $\delta < \delta_0$

$$\alpha(\Delta(z_0,\delta) \cap E) \leq c\gamma_C(\Delta(z_0,r\delta) \cap E) .$$

Exercise 2.6: There is no constant K such that if E has diameter one, $\gamma_C(E) \leq KC(E)$. Let $E_n = [0,1] \times [0,1/n]$. Show $\lim \gamma_C(E_n) = \lim \alpha(E_n) = 1/4$, but $\lim C(E_n) = C([0,1]) = 0$. It may help to observe that if $\{\sigma_n\}$ is a sequence of positive measures converging weak star to σ, then $U_\sigma \leq \underline{\lim}\, U_{\sigma_n}$; and hence Newtonian capacity is "continuous from above."

Exercise 2.7: (a) Show that if $B(E)$ or $B_h(E)$ is non-trivial then it separates the points of S^2.

(b) Show that the maximal ideal space of the algebra $B(E)$ or $B_h(E)$ coincides with S^2 whenever the algebra is non-trivial.

Pages 28-31 of [28] will help here.

Exercise 2.8: In [31] it is proved that if $\Omega = S^2 \setminus E$, $A(\Omega)$ is a dirichlet algebra on E if and only if E is connected and there is $\delta_0 > 0$ and $c > 0$ such that for all $z \in E$, $\delta < \delta_0$

$$\alpha(\Delta(z,\delta) \cap E) > c\delta .$$

Show that $B(E)$ is a dirichlet algebra on E if and only if E is connected and there is δ_0 on C such that if $z \in E$ and $\delta < \delta_0$

$$\gamma_C(\Delta(z_0,\delta) \cap E) \geq C\delta \ .$$

Formulate and prove a theorem characterizing those E for which $B_h(E)$ is a dirichlet algebra on E. (Hint: It is not necessary to read [31] to do the exercise.)

Exercise 2.9: Let E be a compact plane set and let $\Omega = S^2 \setminus E$. Prove the following

(a) There is a constant M such that if $f \in H^\infty(\Omega)$ there is a sequence $\{f_n\}_{n=1}^\infty$ in $B(E)$ such that $\|f_n\| \leq M\|f\|$ and $f_n(z) \to f(z)$ for all $z \in \Omega$ if and only if there is a constant K such that for all $z \in E$ and all $\delta > 0$

$$\gamma(E \cap \Delta(z,\delta)) \leq K\gamma_C(E \cap \Delta(z,2\delta)) \ .$$

See [30, §2].

(b) If the assertions in (a) hold, show one can find $\{f_n\}$ in $B(E)$ such that $\|f_n\| \leq \|f\|$ and $f_n(z) \to f(z)$ on Ω. See [19].

(c) Formulate and prove similar assertions for $B_h(E)$.

(This is a rather lengthy exercise; it consists of reproving in the present context the results in the cited references.)

§3. Some Consequences

We begin by comparing m_h and γ_h. Again let $h(t)$ be a measure function satisfying (*) and let

$$\Phi(\delta) = \int_0^\delta \frac{h(t)}{t^2}\,dt + \frac{h(\delta)}{\delta} = \int_0^\delta \frac{dh(t)}{t} .$$

This function made an appearance in §4 of III, though it was unnamed there.

Lemma 3.1: If E is an analytic set of diameter δ, then

$$m_h(E) \leq 225\Phi(\delta)\gamma_h(E) .$$

Proof: We can assume E is compact. By III 1.5 there is a positive measure σ on E such that $\sigma(\Delta(z,r)) \leq 25h(r)$ for all discs $\Delta(z,r)$ and $\sigma(E) \geq m_h(E)$. Then $\hat{\sigma} \in B_h(E)$, and in III §4 we showed that if we write $h(R) = m_h(E)$, then $\|\hat{\sigma}\| \leq 25\Phi(R)$. Since $m_h(E) \leq 9h(\delta)$, $\Phi(R) \leq 9\Phi(\delta)$, and thus

$$m_h(E) \leq |\hat{\sigma}'(\infty)| \leq 225\Phi(\delta)\gamma_h(E) .$$

Theorem 3.2: Let f be continuous on S^2 and analytic off some compact set E. Set $h(t) = t\omega_f(t)$ and assume $\int h(t)/t^2\,dt < \infty$. Then $f \in B_h(E)$.

Proof: Let S be a square of side δ. By III 2.2 and 3.1 above

$$\left| \int_{\partial S} f(z)dz \right| \leq 4m_h(S \cap E) \leq 4c\Phi(\delta)\gamma_h(S \cap E) ,$$

so that $f \in B_h(E)$ by Theorem 2.4.

Corollary 3.3: (Davie). Let f be continuous on S^2 and analytic on some open set U. Assume $h(t) = t\omega_f(t)$ satisfies $\int h(t)/t^2\,dt < \infty$. If $m_h(L) = 0$, then f can be uniformly approximated by continuous functions analytic on $U \cup L$.

<u>Proof</u>: This follows from 3.2 and 1.3.

Davie's proof of 3.3 in [18] is much more elementary than the one above, but it does not give 3.2. Notice that 3.2 strengthens Theorem III 2.3 significantly.

For functions satisfying a Hölder condition with exponent β, $0 < \beta < 1$, the converse of 3.3 holds as well, because by III 4.5, $\hat{\mu}$ is in Lip_β if μ has growth $t^{1+\beta}$. Write

$$A_\beta(E) = \{f : f \in C(E,1),\ \omega_f(\delta) \leq C\delta^\beta\} .$$

<u>Corollary 3.4</u>: Let $0 < \beta < 1$. If $m_{1+\beta}(L) = 0$ then for any set E, $A_\beta(E \setminus L)$ is uniformly dense in $A_\beta(E)$. Conversely, if for every E, $A_\beta(E \setminus L)$ is uniformly dense in $A_\beta(E)$, then $\underline{m_{1+\beta}}(L) = 0$.

<u>Proof</u>: The first half follows 3.3 and the fact that by III 4.5, $A_\beta(E)$ is dense in $B_h(E)$ if $h(t) = t^{1+\beta}$. For the converse notice that if $\underline{m_{1+\beta}}(L) > 0$, then there is a non-constant function in $A_\beta(L)$ by III 4.5. But if for instance $E \setminus L$ is a singleton, then $A_\beta(E \setminus L)$ is trivial and therefore not dense in $A_\beta(E)$. For $\beta = 1$, the first half of the above theorem is still true, and easy (see Exercise 2.7 of II). Very little is known concerning the converse statement.

<u>Exercise 3.5</u>: Assume $h(t)/t$ is increasing but $\int_0 h(t)/t^2\, dt = \infty$. Define γ_h in terms of continuous Cauchy transforms of measures of growth $h(t)$. Show there exists sets E_δ of diameter δ such that

$$\overline{\lim_{\delta \to 0}} \frac{m_h(E_\delta)}{\gamma_h(E_\delta)} = \infty .$$

This should explain the importance of (*) in the above results.

<u>Exercise 3.6</u>: Let $0 < \beta < 1$. If $A_\beta(E)$ is non-trivial, then $A_\beta(E)$ is not uniformly closed. See Exercise III 3.8.

§4. Rational Approximation

Let K be a compact plane set. Denote by $A(K)$ the functions in $C(K)$ which are analytic on the interior K^0 of K. Thus $A(K)$ consists of all extensions to K of the functions in $A(K^0)$. Let $R(K)$ be the uniform closure in $C(K)$ of the set of functions analytic on a neighborhood of K. Then $R(K) \subset A(K)$.

It is clear that $B(\mathbb{C}\setminus K^0) \subset A(K)$. Fixing a measure function $h(t)$ satisfying the integrability condition (*), we also have $B_h(\mathbb{C}\setminus K^0) \subset A(K)$. Moreover, Theorem 2.5 describes those K for which $B(\mathbb{C}\setminus K^0) = A(K)$ or $B_h(\mathbb{C}\setminus K^0) = A(K)$. We want to determine when $B_h(\mathbb{C}\setminus K^0) = R(K)$ and when $B(\mathbb{C}\setminus K^0) = R(K)$.

We begin with two trivial observations. First, $R(K) \subset B_h(\mathbb{C}\setminus K^0)$, and hence $R(K) \subset B(\mathbb{C}\setminus K^0)$, because $R(K)$ is the closed linear span of the functions $g(z) = (z - a)^{-1}$, $a \notin K$ (Runge's theorem). Such a function is the Cauchy transform of a point mass. If that mass is smeared out into, say, $\mu_\delta = dxdy/\pi\delta^2$ on $\Delta(a,\delta)$, then $\hat{\mu}_\delta$ converges to g uniformly on K. Secondly, the Cauchy transform of any measure with closed support disjoint from K lies in $R(K)$. So our problem is: When is (the restriction to K of) $B(\partial K)$ or $B_h(\partial K)$ contained in $R(K)$?

Theorem 1.1 gives one sufficient condition. The <u>inner boundary</u> of K is $\partial_i K = \partial K \setminus \bigcup_j \partial H_j$ where $\{H_j\}$ are the components of $\mathbb{C}\setminus K$.

It is known [81] that any function in $A(K)$ which has an analytic extension to a neighborhood of $\partial_i K$ is in $R(K)$. With 1.1 this yields

Theorem 4.1: If $m_h(\partial_i K) = 0$, then $B_h(\partial K) \subset R(K)$. If $C(\partial_i K) = 0$, then $B(\partial K) \subset R(K)$.

There is a "softer" proof of 4.1 which we now give. A point $p \in K$ is a peak point for $R(K)$ if there is $f \in R(K)$ such that $f(p) = 1$ and $|f(z)| < 1$ for $z \in K \setminus \{p\}$. It is known that if p is a peak point for $R(K)$ and μ is a measure on K annihilating $R(K)$ then $\hat{\mu}(p) = 0$ whenever $U_{|\mu|}(p) < \infty$. See [28], [86]. This fact yields another criterion for the containments $B_h(E) \subset R(K)$ and $B(E) \subset R(K)$.

Theorem 4.2: Let K be a compact plane subset and let E be a Borel subset of ∂K. Then $B_h(E) \subset R(K)$ if and only if m_h almost all points of E are peak points for $R(K)$. Also, $B(E) \subset R(K)$ if and only if the set of non peak points for $R(K)$ lying in E has Newtonian capacity zero.

Proof: If ν is a measure of growth $h(t)$ and μ is any other measure then

$$\int \frac{d|\mu|(\zeta)d|\nu|(z)}{|\zeta - z|} < \infty$$

because $U_{|\nu|}$ is continuous. So if ν is supported on E and m_h almost all points of E are peak points then for any measure μ annihilating $R(K)$,

$$\int \hat{\nu}(z)d\mu(z) = -\int \hat{\mu}(\zeta)d\nu(\zeta) = 0$$

and $\hat{\nu} \in R(K)$. Conversely if F is any compact subset of E with $m_h(F) > 0$, then $B_h(F) \neq C$, some function in $B_h(F)$ peaks (i.e. attains its norm) on a subset of F. As $B_h(F) \subset R(K)$, this subset is a peak set for $R(K)$, and by a general result of Bishop [8] it contains a peak point. The proof for $B(E)$ is the same.

It is a theorem of Curtis [16], [28], that the point $z_0 \in \partial K$ is a peak point for $R(K)$ if

$$(4.1) \qquad \overline{\lim_{\delta\to 0}} \frac{\gamma(\Delta(z_0,\delta)\setminus K)}{\delta} > 0 .$$

<u>Corollary 4.3</u>: Let K be a compact plane set and let E be a Borel subset of ∂K. If (4.1) holds m_h almost everywhere on E, then $B_h(E) \subset R(K)$. If (4.1) holds for all $z_0 \in E$ except for a set of Newtonian capacity zero, then $B(E) \subset R(K)$.

Theorem 4.1 follows from 4.3 because if $z_0 \in \partial K\setminus\partial_i K$, then by I 1.3, z_0 satisfies (4.1).

We shall now use some ideas from [58] and [81] to get a sharper form of 4.3 in the case of $B_h(E)$. We begin with a metric density theorem and a Vitali covering theorem, both of which are valid for any Hausdorff measure. Neither is sharp, but they suffice for our purposes. For a finer metric density theorem see for instance, [55].

<u>Lemma 4.4</u>: Let h be a measure function and let E be a Borel set. Then

$$\overline{\lim_{\delta\to 0}} \frac{m_h(\Delta(z,\delta) \cap E)}{h(\delta)} \geq \frac{1}{25}$$

at m_h-almost all points of E.

<u>Proof</u>: The set of $z \in E$ where the assertion fails is a Borel set. Assuming the lemma is false, we can find a compact set $F \subset E$ such that

$$\varlimsup_{\delta\to 0} \frac{m_h(\Delta(z,\delta) \cap F)}{h(\delta)} \leq a < \frac{1}{25} \quad \text{for all } z$$

and $m_h(F) > 0$. By Theorem 1.5 of III there is a positive mass σ of growth $h(t)$ supported on F such that $\sigma(F) \geq m_h(F)/25$. If S is a square of side 2^{-n}, then

$$\sigma(S) \leq m_h(S \cap F) \leq ah(2^{-n}) .$$

Cover F by squares S_j from our special grid $\mathcal{G}$ such that

$$\sum h(\delta_j) < (1 + \varepsilon)m_h(F) .$$

Then

$$m_h(F)/25 \leq \sigma(F) \leq \sum \sigma(S_j) \leq a \sum h(\delta_j)$$

$$\leq (1 + \varepsilon)am_h(F)$$

and this is a contradiction if ε is small.

By the <u>double</u> $\tilde{\Delta}$ of a disc $\Delta = \Delta(z,\delta)$ we mean $\Delta(z,2\delta)$.

<u>Lemma 4.5</u>: Let $h(t)$ be a measure function, and let $\{\Delta_j\} = \{\Delta(z_j,\delta_j)\}$ be a sequence of discs such that $\Sigma\, h(\delta_j) = A < \infty$. Then there is a subsequence $\{\Delta_{j_n}\}$ such that

(i) $\tilde{\Delta}_{j_n} \cap \tilde{\Delta}_{j_m} = \emptyset$

(ii) For any disc $\Delta(z,\delta)$

$$\sum_{\Delta_{j_n} \subset \Delta} h(\delta_{j_n}) \leq 2h(\delta)$$

(iii) $\Sigma\, h(\delta_{j_n}) \geq cA$ where c is a universal constant.

Proof: We may assume $\delta_1 \geq \delta_2 \geq \cdots$. Let $\Delta_{j_1} = \Delta_1$ and assuming $\Delta_{j_1},\ldots,\Delta_{j_n}$ have been chosen satisfying (i) and (ii). Take j_{n+1} to be the least index such that $\{\Delta_{j_1},\ldots,\Delta_{j_n},\Delta_{j_{n+1}}\}$ satisfies (i) and (ii). In this way we generate a subsequence satisfying (i) and (ii).

Now if Δ_j is not in the subsequence $\{\Delta_{j_n}\}$ then either

(a) for some $j_m < j$, $\tilde{\Delta}_{j_m} \cap \tilde{\Delta}_j \neq \emptyset$ or

(b) there is a disc Δ_j^* containing Δ_j such that

$$h(\delta_j) + \sum_{\substack{\Delta_{j_n} \subset \Delta_j^* \\ j_n < j}} h(\delta_{j_n}) \geq 2h(\delta_j^*)$$

where Δ_j^* has radius δ_j^*.

Let $\Sigma_j^{(a)}$ denote the sum over those indices for which (a) holds. Since $\delta_{j_m} \geq \delta_j$ if $j_m < j$, we have

$$\sum_j^{(a)} h(\delta_j) \leq c_1 \sum_n h(\delta_{j_n})$$

because c_1 copies of Δ_{j_m} cover all Δ_j, $j > j_m$ for which $\tilde{\Delta}_j \cap \tilde{\Delta}_{j_m} \neq \emptyset$.

Let $\Sigma_j^{(b)}$ denote the sum over those indices for which (b) holds but (a) fails. We can extract a subsequence Δ_k^* of $\{\Delta_j^* : \Delta_j$ has (b)$\}$ such that the doubles of the Δ_k^* are pairwise disjoint. Then as with $\Sigma^{(a)}$, we have

$$\sum^{(b)} h(\delta_j^*) \leq c_1 \sum h(\delta_k^*) .$$

But then by (b),

$$\sum^{(b)} h(\delta_j) \leq \sum^{(b)} h(\delta_j^*)$$

$$\leq c_1 \sum h(\delta_k^*) \leq 2c_1 \sum h(\delta_k)$$

$$+ 2c_1 \sum_k \sum_{\substack{\Delta_{j_n} \subset \Delta_k^* \\ j_n \leq k}} h(\delta_{j_n}) .$$

In the last expression the first sum is dominated by the second, and the second is dominated by $2c_1 \Sigma h(\delta_{j_n})$. Hence

$$\sum h(\delta_j) \leq (1 + 5c_1) \sum h(\delta_{j_n}) .$$

We return to a fixed measure function $h(t)$ with (*) and recall the auxillary function

$$\Phi(\delta) = \int_0^\delta \frac{dh(t)}{t} .$$

<u>Lemma 4.6</u>: Let K be a compact plane set and let E be the set of

$z \in \partial K$ such that

$$\overline{\lim_{\delta\to 0}} \frac{\gamma(\Delta(z,\delta)\setminus K)}{h(\delta)} = \infty .$$

Then there is a universal constant c such that for m_h almost all $z \in E$

$$\underline{\lim_{\delta\to 0}} \frac{\Phi(\delta)\gamma(\Delta(z,\delta)\setminus K)}{h(\delta)} \geq c .$$

Proof: The set E is easily seen to be a Borel set, so that by 4.4, there is a set $E^* \subset E$ such that $m_h(E\setminus E^*) = 0$ and for each $z_0 \in E^*$ there is $\delta(z_0)$ such that

$$\frac{m_h(\Delta(z_0,\delta) \cap E)}{h(\delta)} > \frac{1}{26}$$

for $\delta < \delta(z_0)$. Fix such a point z_0, and fix $\delta < \delta(z_0)$. Let $\eta > 0$ and cover $\Delta(z_0,\eta\delta) \cap E^*$ by discs $\Delta_j = \Delta(z_j,\eta_j)$ such that $z_j \in E^*$, $\gamma(\Delta_j\setminus K) > h(\delta_j)/\Phi(\delta)$, $\delta_j < (1-\eta)\delta$, and $m_h(\Delta(z_0,\eta\delta) \cap E^*) \leq \Sigma\, h(\delta_j) < \infty$. Using 4.5 we extract a subsequence $\{\Delta_k\}$ of these discs see that

(i) $\tilde{\Delta}_k \cap \tilde{\Delta}_m = \emptyset$, if $k \neq m$

(ii) For any disc Δ of radius ρ

$$\sum_{\Delta_k\subset\Delta} h(\delta_k) \leq 2h(\rho) ,$$

(iii) $\sum h(\delta_k) \geq c_1 m_h(\Delta(z_0,\eta\delta) \cap E^*) \geq c_2 h(\eta\delta)$.

Let L_k be a compact subset of $\Delta_k\setminus K$ such that $\gamma(L_k) = h(\delta_k)/\Phi(\delta)$

and let φ_k be the Ahlfors function of L_k. Then for any z,

$$|\Phi(\delta)\varphi_k(z)| \le \frac{h(\delta_k)}{\operatorname{dist}(z,L_k)}$$

for all but one value of k, by condition (i). Let ν be a measure with mass $h(\delta_k)$ on Δ_k; by (ii) ν is of growth $2h(t)$. Hence

$$\Phi(\delta) \sum |\varphi_k(z)| \le \Phi(\delta) + \Phi(\delta) \sum_{\operatorname{dist}(z,L_k)>\delta} \frac{h(\delta_k)}{\operatorname{dist}(z,L_k)}$$

$$\le \Phi(\delta)\left(1 + \int_{\Delta(z_0,\delta)} \frac{d\nu(\zeta)}{|\zeta - z|}\right) \le \Phi(\delta)(1 + 2\Phi(\delta)) .$$

Therefore $\Phi(\delta) \Sigma \varphi_k \in A(\Delta(z_0,\delta)\setminus K, 2\Phi(\delta))$ if δ is sufficiently small, and as $\Sigma \varphi_k'(\infty) \ge \Sigma h(\delta_k) \ge c_2 h(\eta\delta)$, we have

$$\gamma(\Delta(z,\delta)\setminus K) \ge c_2 h(\eta\delta)/2\Phi(\delta) .$$

Sending η to 1 now gives the result.

The argument above is used by Melnikov [58] to approximate Hölder continuous functions (see Theorem 4.9 below). It is also the first step of Vitushkin's proof of the instability of analytic capacity: For any set G area almost all points of the plane satisfy one of the following

$$\lim_{\delta\to 0} \frac{\alpha(G \cap \Delta(z,\delta))}{\delta^2} = 0$$

$$\lim_{\delta\to 0} \frac{\alpha(G \cap \Delta(z,\delta))}{\delta} = 1 .$$

The reasoning on pp. 188-190 of [81] shows that the hypothesis of 4.6 can be weakened to

$$\overline{\lim_{\delta\to 0}} \frac{\gamma(\Delta(z,\delta)\setminus K)}{h(\delta)} > 0 .$$

The conditions in 4.7 and its followers below can be weakened similarly. The reader is referred to [81] for the details, which are similar to the proof of 4.6 but somewhat more technical.

Theorem 4.7: Let K be a compact plane set and let E be a Borel subset of ∂K. Then the following are equivalent

(a) $B_h(E) \subset R(K)$

(b) There is a universal constant c such that

$$\underline{\lim_{\delta\to 0}} \frac{\Phi(\delta)\gamma(\Delta(z,\delta)\setminus K)}{h(\delta)} \geq c ,$$

for m_h almost all points of E.

(c) For m_h almost all points of E,

$$\overline{\lim_{\delta\to 0}} \frac{\gamma(\Delta(z,\delta)\setminus K)}{h(\delta)} = \infty .$$

(d) For every disc $\Delta(z,\delta)$,

$$\gamma_h(E \cap \Delta(z,\delta)) \leq \gamma(\Delta(z,\delta)\setminus K) .$$

Proof: Assume (a) holds. Let $\varepsilon > 0$ and let J be a compact subset of $E \cap \Delta(z,\delta)$ and let $f \in B_h(J)$ such that $\|f\| \leq 1$ and $f'(\infty) = (1 - \varepsilon)\gamma (E \cap \Delta(z,\delta))$. Then f is in $R(K)$ and analytic off $\overline{\Delta(z,\eta\delta)}$

for some $\eta < 1$. It follows that $f \in R(K \cup (\mathbb{C}\setminus\overline{\Delta(z,\delta)}))$, by the localization theorem for rational approximation [28, p. 51]. But this means $|f'(\infty)| \leq \gamma(\Delta(z,\delta)\setminus K)$, so that (a) implies (d).

By Lemma 3.2, (d) implies that (c) (in fact (b)) holds at every point of density of E, and Lemma 4.6 asserts that (c) implies (b).

Now assume (b). Let ν be a measure of growth $h(t)$ supported on E. We must show $\hat{\nu} \in R(K)$. Let E_n be the set of z in E such that

$$\frac{\Phi(\delta)\gamma(\Delta(z,\delta)\setminus K)}{h(\delta)} \geq c/2 \quad \text{for all} \quad \delta < 1/n .$$

Each set E_n is Borel and $m_h(E\setminus E_n) \to 0$ $(n \to \infty)$. So if $\nu_n = \nu|E_n$ then $\hat{\nu}_n$ converges to $\hat{\nu}$ uniformly, and we only have to prove $\hat{\nu}_n \in R(K)$. Let S be any square of side 2δ. Then

$$\left| \int_{\partial S} \hat{\nu}_n(z)dz \right| = \nu_n(S) \leq c_2 h(\delta)$$
$$\leq c_3\Phi(\delta)\gamma(S\setminus K) .$$

By a theorem of Vitushkin [81] analogous to 2.4 above, $\hat{\nu}_n \in R(K)$. Of course we could just as well have used the analogue of 2.3.

Combining 4.7 and 4.2, we have:

Corollary 4.8: Let K be a compact plane set and let E be the set of $z \in \partial K$ for which

$$\overline{\lim_{\delta\to 0}} \frac{\gamma(\Delta(z,\delta)\setminus K)}{h(\delta)} = \infty .$$

Then m_h almost all points of E are peak points for $R(K)$.

When $h(\delta)/\Phi(\delta) = O(\delta)$, or equivalently, when $\int_0 h(t)/(t^{2+\varepsilon})dt < \infty$ for some $\varepsilon > 0$, Theorem 4.8 can be proved without Vitushkin's approximation theorem. For in that case condition (b) of 4.7 is just (4.1), so that m_h almost all points are peak points and 4.2 can be applied. The measure functions $h(t) = t^{1+\beta}$ satisfy the above condition.

Using the results of §3 we obtain two more Corollaries of 4.7.

<u>Corollary 4.9</u>: Let K be a compact set and let $f \in A(K)$. Assume $\int \frac{h(t)}{t^2} dt < \infty$ where $h(t) = t\omega_f(t)$. Then $f \in R(K)$ if either

(i) $\overline{\lim_{\delta\to 0}} \frac{\gamma(\Delta(z,\delta)\setminus K)}{h(\delta)} = \infty$ for m_h almost all points of ∂K.

(ii) m_h almost all points of ∂K are peak points.

<u>Corollary 4.10</u>: (Melnikov [58]). Let $0 < \beta \leq 1$, and let $A_\beta(K)$ be $A(K) \cap \mathrm{Lip}_\beta$. Then the following are equivalent

(i) $A_\beta(K) \subset R(K)$

(ii) there is a constant c such that $\underline{\lim_{\delta\to 0}} \frac{\gamma(\Delta(z,\delta)\setminus K)}{\delta} \geq c$ for $m_{1+\beta}$ almost all points of ∂K.

(iii) $\overline{\lim} \frac{\gamma(\Delta(z,\delta)\setminus K)}{\delta^{1+\beta}} = \infty$ for $m_{1+\beta}$ almost all points of ∂K.

(iv) $m_{1+\beta}$ almost all points of ∂K are peak points for $R(K)$.

It would be interesting to have a version of 4.9 without the hypothesis $\int \frac{\omega_f(t)}{t} dt < \infty$. Vitushkin's characterization of $R(K)$ easily implies that $f \in A(K)$ is in $R(K)$ if there is $\Omega(\delta) \to 0$ $(\delta \to 0)$ such that for all $z \in \partial K$

$$\omega_f(\delta)\alpha(\Delta(z,\delta)\setminus K^0) \leq \Omega(\delta)\gamma(\Delta(z,\delta)\setminus K) .$$

Corollary 2.2 of [29] provides the weaker condition

$$\omega_f(\delta)\alpha(\Delta(z,\delta)\setminus K^0) \leq c\gamma(\Delta(z,\delta)\setminus K) .$$

This can be replaced by

$$\varliminf_{\delta\to 0} \frac{\gamma(\Delta(z,\delta)\setminus K)}{\delta\omega_f(\delta)} > 0$$

for each $z \in \partial K$. However it is not clear that this can be weakened to $\varlimsup_{\delta\to 0} = \infty$. Finally, without the integrability condition (*), (ii) of 4.9 is not sufficient, for Davie [17] has an example of a compact set K for which $R(K) \neq A(K)$ yet each point of ∂K is a peak point for $R(K)$.

Exercise 4.11: Give capacity conditions necessary and sufficient for $B(\partial K) \subset R(K)$.

BIBLIOGRAPHY

This is not a complete list, but more of a sampling. Some excellent related papers can be found in the references bibliographies.

[1] Ahlfors, L. "Bounded analytic functions", Duke Math. J. 14 (1947), 1-11.

[2] Ahlfors, L. Complex Analysis 2nd. ed. McGraw-Hill, New York, 1966.

[3] Ahlfors, L. Extremal Problems in der Funktiontheorie", Ann. Acad. Sci. Fenn. Ser. AI 249/1, (1958).

[4] Ahlfors, L. "Open Riemann surfaces and extremal problems on compact subregions", Comment. Math. Helv. 24 (1950), 100-134.

[5] Ahlfors, L. and Beurling, A. "Conformal invarients and function theoretic null sets", Acta Math. 83 (1950), 101-129.

[6] Arens, R. "The maximal ideals of certain function algebras", Pac. J. Math. 8 (1958), 641-648.

[7] Besicovitch, A. "On sufficient conditions for a function to be analytic and on behavior of analytic functions in the neighborhood of non-isolated singular points", Proc. London Math. Soc. (2) 32 (1931), 1-9.

[8] Bishop, E. "A minimal boundary for function algebras", Pac. J. Math. 9 (1959), 629-642.

[9] Bishop, E. "Approximation by a polynomial and its derivitive on certain closed sets", Proc. Amer. Math. Soc. 9 (1958) 946-953.

[10] Brelot, M. "Élements de la théorie classique du potentel", Les Cours de Sorbonne 3e cycle. 2nd ed. Centre de Documentation Universitaire, Paris (1961).

[11] Browder, A. Introduction to Function Algebras, Benjamin, New York, 1969.

[12] Carleson, L. "On null-sets for continuous analytic functions", Arkiv för Mat. 1 (1950), 311-318.

[13] Carleson, L. "On the connection between Hausdorff measures and capacity", Ark. Mat. 3 (1958), 403-406.

[14] Carleson, L. Selected Problems on Exceptional Sets, Van Nostrand, Princeton, N. J., 1967.

[15] Collingwood, E. and Lohwater, A. The Theory of Cluster Sets, Cambridge University Press, 1966.

[16] Curtis, P. "Peak points for algebras of analytic functions", J. Funct. Anal. 3 (1969), 35-47.

[17] Davie, A. "An example on rational approximation", Bull. London Math. Soc., 2 (1970), 83-86.

[18] Davie, A. "Analytic capacity and approximation problems" Trans. Amer. Math. Soc. (to appear).

[19] Davie, A. "Bounded approximation and dirichlet sets", J. Funct. Anal. 6 (1970), 460-467.

[20] Davie, A. and Wilkin, D. "An extension of Melnikov's theorem", to appear.

[21] Denjoy, A. "Sur la continuité des fonctions analytiques singulières", Bull. Soc. Math. France 60 (1932), 27-105.

[22] Denjoy, A. "Sur les fonctions analytiques uniformes à singularités discontinues", C. R. Acad. Sci. Paris, 149 (1909), 258-260.

[23] Dolzhenko, E. P. "The removability of singularities of analytic functions", Uspehi Mat. Nauk., 18 (1963) (112), 135-142 (Russian).

[24] Dunford, N., and Schwartz, J. Linear Operators, Part I. Interscience, New York 1958.

[25] Duren, P. Theory of H^p Spaces, Academic Press, New York 1970.

[26] Fisher, S. "On Schwarz's lemma and inner functions", Trans. Amer. Math. Soc. 138 (1969), 229-240.

[27] Frostman, O. "Potential d'équilibre et capacité des ensembles avec quelques applications à la théorie des fonctions. Meddel. Lunds Univ. Mat. Sem. 3 (1935), 1-118.

[28] Gamelin, T. Uniform Algebras, Prentice-Hall, Englewood Cliffs N. J., 1969.

[29] Gamelin, T. and Garnett, J. "Bounded approximation by rational functions", Pac. J. Math. (to appear).

[30] Gamelin, T. and Garnett, J. "Constructive techniques in rational approximation", Trans. Amer. Math. Soc. 143 (1969), 187-200.

[31] Gamelin T. and Garnett, J. "Pointwise bounded approximation and dirichlet algebras", J. Funct. Anal. (to appear).

[32] Garabedian, P. "Schwarz's lemma and the Szëgo kernel function", Trans. Amer. Mat. Soc. 67 (1949), 1-35.

[33] Garnett, J. "Metric conditions for rational approximation", Duke Math. J. 37 (1970), 689-699.

[34] Garnett, J. "Positive length but zero analytic capacity", Proc. Amer. Math. Soc. 21 (1970), 696-699.

[35] Glicksberg, I. "The abstract F. and M. Riesz theorem", J. Funct. Anal. 1 (1967), 109-122.

[36] Goluzin, G. M. Geometric Theory of Functions of a Complex Variable, Amer. Math. Soc., Providence, R. I. 1969.

[37] Hallstrom, A. "Bounded Point Derivations and Other Topics concerning Rational Approximation", Thesis, Brown U. 1968.

[38] Halmos, P. Measure Theory, Van Nostrand, New York, 1950.

[39] Havin, V. P. "Approximation in the mean by analytic functions" Soviet. Math Dokl. 6 (1965), 1458-1460.

[40] Havin, V. P. "On analytic functions representable by an integral of Cauchy-Stieltjes type", Vestnik Lenngrad. Univ. ser. mat. meh. astron. 13 (1958), 66-79, (Russian).

[41] Havin, V. P. "On the space of bounded regular functions", Sibrisk. Mat. Zh. 2 (1961), 622-638. (Russian).

[42] Havin, V. P. "On the space of bounded regular functions", Soviet Math. Dokl. 1 (1960), 202-204.

[43] Havin, V P. "Spaces of analytic functions", Mathematical Analysis 1964. Akad-Nauk SSSR Inst. Naučn. Informacci, Moscow, 1966, 76-1164 (Russian) M.R. 34 #6512

[44] Havin, V. P. and Havinson S. Ja. "Some estimates of analytic capacity", Soviet Math. Dokl. 2 (1961), 731-734.

[45] Havinson, S. Ya. "Analytic capacity of sets, joint nontriviality of various classes of analytic functions and the Schwarz lemma in arbitrary domains", Amer. Math. Soc. Translations, ser. 2 vol. 43, 215-266.

[46] Havinson, S. Ja. "The theory of extremal problems for bounded analytic functions satisfying additional conditions inside the domain", Uspehi Mat. Nauk 18 (1963) vol. 2 (110), 25-68. (Russian).

[47] Hedberg, L. "Approximation in the mean by analytic functions", Trans. Amer. Math. Soc. (to appear).

[48] Hedberg, L. "Approximation in the mean by analytic and harmonic functions, and capacities", to appear.

[49] Helms, L. Introduction to Potential Theory, Wiley-Interscience, New York 1969.

[50] Hoffman, K. Banach Spaces of Analytic Functions, Prentice-Hall, Englewood Cliffs, N.J. 1962.

[51] Hörmander, L. Linear Partial Differential Operators, Springer, Berlin, 1963.

[52] Ivanov, L. D. "On Denjoy's conjecture", Uspehi Mat. Nauk. 18 (112) (1963), 147-149. (Russian).

[53] Ivanov, L. D. "On the analytic capacity of linear sets", Uspehi Mat. Nauk. 17 (108) (1962), 143-144. (Russian).

[54] Kametani, S. "On Hausdorff's measures and generalized capacities with some applications to the theory of functions", Jap. Jour. Math. 19 (1945), 217-257.

[55] Kametani, S. "On some properties of Hausdorff's measure and the concept of capacity in generalized potentials", Proc. Imp. Acad. Tokyo 18 (1942), 617-675.

[56] Melnikov, M. S. "A bound for the Cauchy integral along an analytic curve", Mat. Sbornik, 71 (113) (1966), 503-515. (Russian). Amer. Math. Soc. Translations ser. 2. 80 (1969), 243-256.

[57] Melnikov, M. S. "Analytic capacity and the Cauchy integral", Soviet Math. Dokl. 8 (1967), 20-23.

[58] Melnikov, M. S. "Metric properties of analytic α-capacity and approximation of analytic functions with Hölder condition by rational functions", Math. Sbornik 8 (1969), 115-124.

[59] Nehari, Z. "Bounded analytic functions", Bull. Amer. Math. Soc. 47 (1951), 354-366.

[60] Nehari, Z. Conformal Mapping, McGraw-Hill, New York, 1962.

[61] Nehari, Z. "On bounded analytic functions", Proc. Amer. Math. Soc. 1 (1950), 268-275.

[62] Nevanlinna, R. Eindeutige Analytische Funktionen, 2nd. ed. Springer, Berlin 1933.

[63] Noshiro, K. "Some remarks on cluster sets", J. Analyse Math. 19 (1967), 283-294.

[64] Ohtsuko, M. "Capacité d' ensembles de Cantor généralisés", Nagoya Math. J. 11 (1957), 151-160.

[65] O'Neill, B. and Wermer, J. "Parts as finite-sheeted coverings of the disc", Amer. J. Math. 90 (1968), 98-107.

[66] Pommerenke, Ch. "Uber die analytische Kapazitat", Arckiv der Math. 11 (1960), 270-277.

[67] Privilov, V. L. Randeigenschaften Analytischer Funktionen, Deutscher Verlag der Wiss., Berlin, 1956.

[68] Riesz, F. and M. "Uber die Randwerte einer analytischen Funktionen" Quatrième Congres des Math. Scand. Stockholm (1916), 27-44.

[69] Rogers, C. Hausdorff Measures, Cambridge University Press 1970.

[70] Roydan, H. "A generalization of Morera's Theorem", Ann. Polon. Math. 12 (1962), 199-202.

[71] Roydan, H. "The boundary values of analytic and harmonic functions", Math. Zeitschr. 78 (1962), 1-24.

[72] Rudin, W. "Analytic functions of class H_p" Trans. Amer. Math. Soc. 78 (1955), 46-66.

[73] Rudin, W. Real and Complex Analysis, McGraw-Hill, New York, 1966.

[74] Schwartz, L. Theorie des distributions I, II, Herman, Paris, 1950-51.

[75] Taylor, S. J. "On connexion between Hausdorff measures and generalized capacity", Proc. Cambs. Philos. Soc. 47 (1961) 524-531.

[76] Trjitzinsky, W. "Problems of representation and uniqueness for functions of a complex variable", Acta Math. 78 (1946), 97-192.

[77] Tsuji, M. Potential Theory in Modern Function Theory, Maruzen Co. Ltd., Tokyo, 1959.

[78] Tumarkin, G. C. "On integrals of Cauchy-Stieltjes type", Uspehi Mat. Nauk. 11 (1956) no. 4 (70) 163-166. (Russian).

[79] Tumarkin, G. C. "Properties of analytic functions representable by integrals of Cauchy-Stieltjes and Cauchy-Lebesgue type. Izv. Akad. Nauk Armjan. SSSR ser. Fiz. - Mat. Nauk 16 (1963), 23-45. (Russian).

[80] Valskii, R.E. "Remarks on bounded functions representable by an integral of the Cauchy-Stieltjes type", Sib. Math. J. 7 (1967), 202-209.

[81] Vitushkin, A. G. "Analytic capacity of sets and problems in approximation theory", Russian Math. Surveys 22 (1967), 139-200.

[82] Vitushkin, A. G. "Estimates of the Cauchy integral", Mat. Sbornik 71 (113) (1966), 515-535. A. M. S. Translations ser. 2. 80 (1969), 257-278.

[83] Vitushkin, A. G. "Example of a set of positive length but of zero analytic capacity", Dokl. Akad. Nauk. SSSR 127 (1959), 246-249. (Russian).

[84] Vitushkin, A. G. "On a problem of Denjoy", Izv. Akad. Nauk SSSR. ser. mat. 28 (1964), 745-756. (Russian).

[85] Wermer, J. "Banach Algebras and analytic functions", Advances in Math. 1 (1961) Fasc. 1, 51-102.

[86] Wilken, D. Lebesgue measure for parts of R(X)", Proc. Amer. Math. Soc. 18 (1967), 508-512.

[87] Wilken, D. "The support of representing measures for R(X)", Pac. Jour. Math. 26 (1968), 621-626.

[88] Zalcman, L. "Analytic Capacity and Rational Approximation", Lecture Notes in Math. No. 50, Springer, Berlin 1968.

[89] Zalcman, L. "Null sets for a class of analytic functions", Amer. Math. Monthly, 75 (1968), 462-470.

[90] Zygmund, A. "Smooth Functions", Duke Math. J. 12 (1945), 47-76.

[91] Zygmund, A. Trigonometric Series, Cambridge Univ. Press, London and New York, 1959.

Author partially supported by NSF grant GP-71475

PB-8480-27
5-12T

Lecture Notes in Mathematics

Comprehensive leaflet on request

Vol. 111: K. H. Mayer, Relationen zwischen charakteristischen Zahlen. III, 99 Seiten. 1969. DM 16,–

Vol. 112: Colloquium on Methods of Optimization. Edited by N. N. Moiseev. IV, 293 pages. 1970. DM 18,–

Vol. 113: R. Wille, Kongruenzklassengeometrien. III, 99 Seiten. 1970. DM 16,–

Vol. 114: H. Jacquet and R. P. Langlands, Automorphic Forms on GL (2). VII, 548 pages. 1970. DM 24,–

Vol. 115: K. H. Roggenkamp and V. Huber-Dyson, Lattices over Orders I. XIX, 290 pages. 1970. DM 18,–

Vol. 116: Séminaire Pierre Lelong (Analyse) Année 1969. IV, 195 pages. 1970. DM 16,–

Vol. 117: Y. Meyer, Nombres de Pisot, Nombres de Salem et Analyse Harmonique. 63 pages. 1970. DM 16,–

Vol. 118: Proceedings of the 15th Scandinavian Congress, Oslo 1968. Edited by K. E. Aubert and W. Ljunggren. IV, 162 pages. 1970. DM 16,–

Vol. 119: M. Raynaud, Faisceaux amples sur les schémas en groupes et les espaces homogènes. III, 219 pages. 1970. DM 16,–

Vol. 120: D. Siefkes, Büchi's Monadic Second Order Successor Arithmetic. XII, 130 Seiten. 1970. DM 16,–

Vol. 121: H. S. Bear, Lectures on Gleason Parts. III, 47 pages. 1970. DM 16,–

Vol. 122: H. Zieschang, E. Vogt und H.-D. Coldewey, Flächen und ebene diskontinuierliche Gruppen. VIII, 203 Seiten. 1970. DM 16,–

Vol. 123: A. V. Jategaonkar, Left Principal Ideal Rings. VI, 145 pages. 1970. DM 16,–

Vol. 124: Séminare de Probabilités IV. Edited by P. A. Meyer. IV, 282 pages. 1970. DM 20,–

Vol. 125: Symposium on Automatic Demonstration. V, 310 pages. 1970. DM 20,–

Vol. 126: P. Schapira, Théorie des Hyperfonctions. XI, 157 pages. 1970. DM 16,–

Vol. 127: I. Stewart, Lie Algebras. IV, 97 pages. 1970. DM 16,–

Vol. 128: M. Takesaki, Tomita's Theory of Modular Hilbert Algebras and its Applications. II, 123 pages. 1970. DM 16,–

Vol. 129: K. H. Hofmann, The Duality of Compact Semigroups and C*- Bigebras. XII, 142 pages. 1970. DM 16,–

Vol. 130: F. Lorenz, Quadratische Formen über Körpern. II, 77 Seiten. 1970. DM 16,–

Vol. 131: A Borel et al., Seminar on Algebraic Groups and Related Finite Groups. VII, 321 pages. 1970. DM 22,–

Vol. 132: Symposium on Optimization. III, 348 pages. 1970. DM 22,–

Vol. 133: F. Topsøe, Topology and Measure. XIV, 79 pages. 1970. DM 16,–

Vol. 134: L. Smith, Lectures on the Eilenberg-Moore Spectral Sequence. VII, 142 pages. 1970. DM 16,–

Vol. 135: W. Stoll, Value Distribution of Holomorphic Maps into Compact Complex Manifolds. II, 267 pages. 1970. DM 18,–

Vol. 136: M. Karoubi et al., Séminaire Heidelberg-Saarbrücken-Strasbourg sur la K-Théorie. IV, 264 pages. 1970. DM 18,–

Vol. 137: Reports of the Midwest Category Seminar IV. Edited by S. MacLane. III, 139 pages. 1970. DM 16,–

Vol. 138: D. Foata et M. Schützenberger, Théorie Géométrique des Polynômes Eulériens. V, 94 pages. 1970. DM 16,–

Vol. 139: A. Badrikian, Séminaire sur les Fonctions Aléatoires Linéaires et les Mesures Cylindriques. VII, 221 pages. 1970. DM 18,–

Vol. 140: Lectures in Modern Analysis and Applications II. Edited by C. T. Taam. VI, 119 pages. 1970. DM 16,–

Vol. 141: G. Jameson, Ordered Linear Spaces. XV, 194 pages. 1970. DM 16,–

Vol. 142: K. W. Roggenkamp, Lattices over Orders II. V, 388 pages. 1970. DM 22,–

Vol. 143: K. W. Gruenberg, Cohomological Topics in Group Theory. XIV, 275 pages. 1970. DM 20,–

Vol. 144: Seminar on Differential Equations and Dynamical Systems, II. Edited by J. A. Yorke. VIII, 268 pages. 1970. DM 20,–

Vol. 145: E. J. Dubuc, Kan Extensions in Enriched Category Theory. XVI, 173 pages. 1970. DM 16,–

Vol. 146: A. B. Altman and S. Kleiman, Introduction to Grothendieck Duality Theory. II, 192 pages. 1970. DM 18,–

Vol. 147: D. E. Dobbs, Cech Cohomological Dimensions for Commutative Rings. VI, 176 pages. 1970. DM 16,–

Vol. 148: R. Azencott, Espaces de Poisson des Groupes Localement Compacts. IX, 141 pages. 1970. DM 16,–

Vol. 149: R. G. Swan and E. G. Evans, K-Theory of Finite Groups and Orders. IV, 237 pages. 1970. DM 20,–

Vol. 150: Heyer, Dualität lokalkompakter Gruppen. XIII, 372 Seiten. 1970. DM 20,–

Vol. 151: M. Demazure et A. Grothendieck, Schémas en Groupes I. (SGA 3). XV, 562 pages. 1970. DM 24,–

Vol. 152: M. Demazure et A. Grothendieck, Schémas en Groupes II. (SGA 3). IX, 654 pages. 1970. DM 24,–

Vol. 153: M. Demazure et A. Grothendieck, Schémas en Groupes III. (SGA 3). VIII, 529 pages. 1970. DM 24,–

Vol. 154: A. Lascoux et M. Berger, Variétés Kähleriennes Compactes. VII, 83 pages. 1970. DM 16,–

Vol. 155: Several Complex Variables I, Maryland 1970. Edited by J. Horváth. IV, 214 pages. 1970. DM 18,–

Vol. 156: R. Hartshorne, Ample Subvarieties of Algebraic Varieties. XIV, 256 pages. 1970. DM 20,–

Vol. 157: T. tom Dieck, K. H. Kamps und D. Puppe, Homotopietheorie. VI, 265 Seiten. 1970. DM 20,–

Vol. 158: T. G. Ostrom, Finite Translation Planes. IV. 112 pages. 1970. DM 16,–

Vol. 159: R. Ansorge und R. Hass. Konvergenz von Differenzenverfahren für lineare und nichtlineare Anfangswertaufgaben. VIII, 145 Seiten. 1970. DM 16,–

Vol. 160: L. Sucheston, Constributions to Ergodic Theory and Probability. VII, 277 pages. 1970. DM 20,–

Vol. 161: J. Stasheff, H-Spaces from a Homotopy Point of View. VI, 95 pages. 1970. DM 16,–

Vol. 162: Harish-Chandra and van Dijk, Harmonic Analysis on Reductive p-adic Groups. IV, 125 pages. 1970. DM 16,–

Vol. 163: P. Deligne, Equations Différentielles à Points Singuliers Reguliers. III, 133 pages. 1970. DM 16,–

Vol. 164: J. P. Ferrier, Seminaire sur les Algebres Complètes. II, 69 pages. 1970. DM 16,–

Vol. 165: J. M. Cohen, Stable Homotopy. V, 194 pages. 1970. DM 16,–

Vol. 166: A. J. Silberger, PGL_2 over the p-adics: its Representations, Spherical Functions, and Fourier Analysis. VII, 202 pages. 1970. DM 18,–

Vol. 167: Lavrentiev, Romanov and Vasiliev, Multidimensional Inverse Problems for Differential Equations. V, 59 pages. 1970. DM 16,–

Vol. 168: F. P. Peterson, The Steenrod Algebra and its Applications: A conference to Celebrate N. E. Steenrod's Sixtieth Birthday. VII, 317 pages. 1970. DM 22,–

Vol. 169: M. Raynaud, Anneaux Locaux Henséliens. V, 129 pages. 1970. DM 16,–

Vol. 170: Lectures in Modern Analysis and Applications III. Edited by C. T. Taam. VI, 213 pages. 1970. DM 18,–

Vol. 171: Set-Valued Mappings, Selections and Topological Properties of 2^X. Edited by W. M. Fleischman. X, 110 pages. 1970. DM 16,–

Vol. 172: Y.-T. Siu and G. Trautmann, Gap-Sheaves and Extension of Coherent Analytic Subsheaves. V, 172 pages. 1971. DM 16,–

Vol. 173: J. N. Mordeson and B. Vinograde, Structure of Arbitrary Purely Inseparable Extension Fields. IV, 138 pages. 1970. DM 16,–

Vol. 174: B. Iversen, Linear Determinants with Applications to the Picard Scheme of a Family of Algebraic Curves. VI, 69 pages. 1970. DM 16,–

Vol. 175: M. Brelot, On Topologies and Boundaries in Potential Theory. VI, 176 pages. 1971. DM 18,–

Vol. 176: H. Popp, Fundamentalgruppen algebraischer Mannigfaltigkeiten. IV, 154 Seiten. 1970. DM 16,–

Vol. 177: J. Lambek, Torsion Theories, Additive Semantics and Rings of Quotients. VI, 94 pages. 1971. DM 16,–

Please turn over

Vol. 178: Th. Bröcker und T. tom Dieck, Kobordismentheorie. XVI, 191 Seiten. 1970. DM 18,–

Vol. 179: Seminaire Bourbaki – vol. 1968/69. Exposés 347-363. IV. 295 pages. 1971. DM 22,–

Vol. 180: Séminaire Bourbaki – vol. 1969/70. Exposés 364-381. IV, 310 pages. 1971. DM 22,–

Vol. 181: F. DeMeyer and E. Ingraham, Separable Algebras over Commutative Rings. V, 157 pages. 1971. DM 16.–

Vol. 182: L. D. Baumert. Cyclic Difference Sets. VI, 166 pages. 1971. DM 16,–

Vol. 183: Analytic Theory of Differential Equations. Edited by P. F. Hsieh and A. W. J. Stoddart. VI, 225 pages. 1971. DM 20,–

Vol. 184: Symposium on Several Complex Variables, Park City, Utah, 1970. Edited by R. M. Brooks. V, 234 pages. 1971. DM 20,–

Vol. 185: Several Complex Variables II, Maryland 1970. Edited by J. Horváth. III, 287 pages. 1971. DM 24,–

Vol. 186: Recent Trends in Graph Theory. Edited by M. Capobianco/J. B. Frechen/M. Krolik. VI, 219 pages. 1971. DM 18.–

Vol. 187: H. S. Shapiro, Topics in Approximation Theory. VIII, 275 pages. 1971. DM 22,–

Vol. 188: Symposium on Semantics of Algorithmic Languages. Edited by E. Engeler. VI, 372 pages. 1971. DM 26,–

Vol. 189: A. Weil, Dirichlet Series and Automorphic Forms. V, 164 pages. 1971. DM 16,–

Vol. 190: Martingales. A Report on a Meeting at Oberwolfach, May 17-23, 1970. Edited by H. Dinges. V, 75 pages. 1971. DM 16,–

Vol. 191: Séminaire de Probabilités V. Edited by P. A. Meyer. IV, 372 pages. 1971. DM 26,–

Vol. 192: Proceedings of Liverpool Singularities – Symposium I. Edited by C. T. C. Wall. V, 319 pages. 1971. DM 24,–

Vol. 193: Symposium on the Theory of Numerical Analysis. Edited by J. Ll. Morris. VI, 152 pages. 1971. DM 16,–

Vol. 194: M. Berger, P. Gauduchon et E. Mazet. Le Spectre d'une Variété Riemannienne. VII, 251 pages. 1971. DM 22,–

Vol. 195: Reports of the Midwest Category Seminar V. Edited by J.W. Gray and S. Mac Lane.III, 255 pages. 1971. DM 22,–

Vol. 196: H-spaces – Neuchâtel (Suisse)- Août 1970. Edited by F. Sigrist, V, 156 pages. 1971. DM 16,–

Vol. 197: Manifolds – Amsterdam 1970. Edited by N. H. Kuiper. V, 231 pages. 1971. DM 20,–

Vol. 198: M. Hervé, Analytic and Plurisubharmonic Functions in Finite and Infinite Dimensional Spaces. VI, 90 pages. 1971. DM 16.–

Vol. 199: Ch. J. Mozzochi, On the Pointwise Convergence of Fourier Series. VII, 87 pages. 1971. DM 16,–

Vol. 200: U. Neri, Singular Integrals. VII, 272 pages. 1971. DM 22,–

Vol. 201: J. H. van Lint, Coding Theory. VII, 136 pages. 1971. DM 16,–

Vol. 202: J. Benedetto, Harmonic Analysis on Totally Disconnected Sets. VIII, 261 pages. 1971. DM 22,–

Vol. 203: D. Knutson, Algebraic Spaces. VI, 261 pages. 1971. DM 22,–

Vol. 204: A. Zygmund, Intégrales Singulières. IV, 53 pages. 1971. DM 16,–

Vol. 205: Séminaire Pierre Lelong (Analyse) Année 1970. VI, 243 pages. 1971. DM 20,–

Vol. 206: Symposium on Differential Equations and Dynamical Systems. Edited by D. Chillingworth. XI, 173 pages. 1971. DM 16,–

Vol. 207: L. Bernstein, The Jacobi-Perron Algorithm – Its Theory and Application. IV, 161 pages. 1971. DM 16,–

Vol. 208: A. Grothendieck and J. P. Murre, The Tame Fundamental Group of a Formal Neighbourhood of a Divisor with Normal Crossings on a Scheme. VIII, 133 pages. 1971. DM 16,–

Vol. 209: Proceedings of Liverpool Singularities Symposium II. Edited by C. T. C. Wall. V, 280 pages. 1971. DM 22,–

Vol. 210: M. Eichler, Projective Varieties and Modular Forms. III, 118 pages. 1971. DM 16,–

Vol. 211: Théorie des Matroïdes. Edité par C. P. Bruter. III, 108 pages. 1971. DM 16,–

Vol. 212: B. Scarpellini, Proof Theory and Intuitionistic Systems. VII, 291 pages. 1971. DM 24,–

Vol. 213: H. Hogbe-Nlend, Théorie des Bornologies et Applications. V, 168 pages. 1971. DM 18,–

Vol. 214: M. Smorodinsky, Ergodic Theory, Entropy. V, 64 pages. 1971. DM 16,–

Vol. 215: P. Antonelli, D. Burghelea and P. J. Kahn, The Concordance-Homotopy Groups of Geometric Automorphism Groups. X, 140 pages. 1971. DM 16,–

Vol. 216: H. Maaß, Siegel's Modular Forms and Dirichlet Series. VII, 328 pages. 1971. DM 20,–

Vol. 217: T. J. Jech, Lectures in Set Theory with Particular Emphasis on the Method of Forcing. V, 137 pages. 1971. DM 16,–

Vol. 218: C. P. Schnorr, Zufälligkeit und Wahrscheinlichkeit. IV, 212 Seiten 1971. DM 20,–

Vol. 219: N. L. Alling and N. Greenleaf, Foundations of the Theory of Klein Surfaces. IX, 117 pages. 1971. DM 16,–

Vol. 220: W. A. Coppel, Disconjugacy. V, 148 pages. 1971. DM 16,–

Vol. 221: P. Gabriel und F. Ulmer, Lokal präsentierbare Kategorien. V, 200 Seiten. 1971. DM 18,–

Vol. 222: C. Meghea, Compactification des Espaces Harmoniques. III, 108 pages. 1971. DM 16,–

Vol. 223: U. Felgner, Models of ZF-Set Theory. VI, 173 pages. 1971. DM 16,–

Vol. 224: Revètements Etales et Groupe Fondamental. (SGA 1). Dirigé par A. Grothendieck XXII, 447 pages. 1971. DM 30,–

Vol. 225: Théorie des Intersections et Théorème de Riemann-Roch. (SGA 6). Dirigé par P. Berthelot, A. Grothendieck et L. Illusie. XII, 700 pages. 1971. DM 40,–

Vol. 226: Seminar on Potential Theory, II. Edited by H. Bauer. IV, 170 pages. 1971. DM 18,–

Vol. 227: H. L. Montgomery, Topics in Multiplicative Number Theory. IX, 178 pages. 1971. DM 18,–

Vol. 228: Conference on Applications of Numerical Analysis. Edited by J. Ll. Morris. X, 358 pages. 1971. DM 26,–

Vol. 229: J. Väisälä, Lectures on n-Dimensional Quasiconformal Mappings. XIV, 144 pages. 1971. DM 16,–

Vol. 230: L. Waelbroeck, Topological Vector Spaces and Algebras. VII, 158 pages. 1971. DM 16,–

Vol. 231: H. Reiter, L^1-Algebras and Segal Algebras. XI, 113 pages. 1971. DM 16,–

Vol. 232: T. H. Ganelius, Tauberian Remainder Theorems. VI, 75 pages. 1971. DM 16,–

Vol. 233: C. P. Tsokos and W. J. Padgett. Random Integral Equations with Applications to Stochastic Systems. VII, 174 pages. 1971. DM 18,–

Vol. 234: A. Andreotti and W. Stoll. Analytic and Algebraic Dependence of Meromorphic Functions. III, 390 pages. 1971. DM 26,–

Vol. 235: Global Differentiable Dynamics. Edited by O. Hájek, A. J. Lohwater, and R. McCann. X, 140 pages. 1971. DM 16,–

Vol. 236: M. Barr, P. A. Grillet, and D. H. van Osdol. Exact Categories and Categories of Sheaves. VII, 239 pages. 1971, DM 20,–

Vol. 237: B. Stenström. Rings and Modules of Quotients. VII, 136 pages. 1971. DM 16,–

Vol. 238: Der kanonische Modul eines Cohen-Macaulay-Rings. Herausgegeben von Jürgen Herzog und Ernst Kunz. VI, 103 Seiten. 1971. DM 16,–

Vol. 239: L. Illusie, Complexe Cotangent et Déformations I. XV, 355 pages. 1971. DM 26,–

Vol. 240: A. Kerber, Representations of Permutation Groups I. VII, 192 pages. 1971. DM 18,–

Vol. 241: S. Kaneyuki, Homogeneous Bounded Domains and Siegel Domains. V, 89 pages. 1971. DM 16,–

Vol. 242: R. R. Coifman et G. Weiss, Analyse Harmonique Non-Commutative sur Certains Espaces. V, 160 pages. 1971. DM 16,–

Vol. 243: Japan-United States Seminar on Ordinary Differential and Functional Equations. Edited by M. Urabe. VIII, 332 pages. 1971. DM 26,–

Vol. 244: Séminaire Bourbaki – vol. 1970/71. Exposés 382–399. IV, 356 pages. 1971. DM 26,–

Vol. 245: D. E. Cohen, Groups of Cohomological Dimension One. V, 99 pages. 1972. DM 16,–